ORGANOMETALLIC CHEMISTRY
An Overview

ORGANOMETALLIC CHEMISTRY
An Overview

John S. Thayer

Dedication

to Darl H. McDaniel

Friend and Colleague,
In grateful appreciation
for all his help and support.

Dr. John S. Thayer
Department of Chemistry
University of Cincinnati
Cincinnati, OH 45221

Library of Congress Cataloging-in-Publication Data

Thayer, John S.
    Organometallic chemistry.

    Bibliography: p.
    Includes index.
    1. Organometallic chemistry.   I. Title.
QD411.T47   1987   547'.05   87-21654
ISBN 0-89573-121-5

©1988 VCH Publishers, Inc.

Printed in the United States of America.

ISBN 0-89573-121-5 VCH Publishers
ISBN 3-527-26196-6 VCH Verlagsgesellschaft

Distributed in North America by:

VCH Publishers, Inc.
220 East 23rd Street, Suite 909
New York, New York 10010

Distributed Worldwide by:

VCH Verlagsgesellschaft mbH
P.O. Box 1260/1280
D-6940 Weinheim
Federal Republic of Germany

Preface

Organometallic chemistry has become one of the most exciting and active areas in the chemical sciences. Reference books and monographs in this area abound; there are even a few textbooks for specialists and aspiring specialists. Standard texts in inorganic or organic chemistry, however, do not cover organometallic chemistry as an entity. They provide detailed discussions of selected portions and ignore the remainder. There is no currently extant book written on organometallic chemistry for the person with a reasonably good background in chemistry who may be interested in learning something about the subject.

This book is written to fill that gap. It gives a survey of current activity in organometallic chemistry, especially in industrial and biological applications. In the interest of brevity, descriptive detail has been kept to a minimum; readers seeking more extensive information will have many reference sources listed to aid them. The author has emphasized topics not readily available elsewhere (eg, organometalloidal chemistry). Topics discussed in standard inorganic or organic textbooks receive less discussion here than their research activity might justify, simply because they are covered elsewhere.

A number of people helped in the preparation of this book. The author thanks Ms. Eliza Mark, whose timely encouragement and interest helped this volume get started, and Dr. Edmund Immergut of VCH Publishers, whose patience at unexpected delays and whose helpfulness helped the volume reach print. The author thanks Dr. Joyce Y. Corey for her meticulous and painstaking review of the preliminary manuscript, and his colleagues at the University of Cincinnati, especially Dr. Darl McDaniel, for their continuing encouragement. Thanks also go to Academic Press for their permission to include some material from my book "Organometallic Compounds and Living Organisms," and to the various editors and workers of VCH Publishers who brought this book into the light of day.

John S. Thayer
Cincinnati, Ohio

Contents

Chapter 5

Metal–Carbon Sigma Bonds. I: The Reactive Metals 39

Chapter 6

Metal–Carbon Sigma Bonds. II: The Heavy Metals 48

Chapter 7

Metal–Carbon Sigma Bonds. III: The Metalloids 62

Chapter 8

Metal–Carbon Synergistic Bonds. I: Mononuclear Compounds of Divalent Carbon 81

Chapter 9

Metal–Carbon Synergistic Bonds. II: Mononuclear Complexes of Unsaturated Hydrocarbons 92

Chapter 10

Metal–Carbon Synergistic Bonds. III: Polynuclear Compounds 102

Chapter 11

Organometallic Compounds in Biology. I: Medicinal and Biochemical Uses 110

Chapter 12

**Organometallic Compounds in Biology. II: Toxicological and Biocidal
Aspects** 122

Chapter 13

**Organometallic Compounds in Biology. III: Environmental
Occurence and Transformation** 132

Chapter 14

Afterword 143

Chapter 1

Organometallic Chemistry: Development of a Discipline

1.1 EVOLUTION OF A CONCEPT

Organometallic chemistry has developed into a recognized subarea of chemistry, having its own practitioners, its own monographs, its own specialized journals, symposia, associations, etc.—all the symbols and stigmata that scientific disciplines and subareas possess. Of all the subdivisions of chemistry, organometallic chemistry is arguably the most diversified, drawing from, and contributing to, virtually every other portion of the chemical sciences, as well as interacting with the physical and biological sciences as well. It is an extremely active field, with thousands of research papers being published each year, conferences proliferating, a growing series of monographs, etc. While most areas of chemistry have prospered in recent decades, organometallic chemistry has flourished better than most.

Almost every element in the Periodic Table can form compounds with carbon. The physical and chemical properties of these compounds will depend very much on the nature of the element bonding to carbon, as was recognized long ago by Kraus[1]:

> Organic chemists have long made use of the organic derivatives of the halogens and of oxygen, sulfur, nitrogen, etc., and most chemists consider that these compounds are typically organic and that their reactions are typical of organic compounds. This, however, is not strictly true. All these compounds have an in-organic side, so to speak; they have properties that are primarily determined by their inorganic constituents, and these properties, in many instances, are the very ones that render these compounds important . . . the chemistry of the organic derivatives of oxygen or nitrogen, for example, cannot be differentiated from that of tin, germanium or boron . . .

The dominating reactions of organic compounds containing elements other than carbon or hydrogen occur at, or arise from, those other elements.

This dominance has become symbolized by the term "functional group"—a term that every student learns in the introductory course in organic chemistry. In this sense, then, organometallic compounds may be regarded as hydrocarbon derivatives having metals as functional groups. This "organic viewpoint" can explain a substantial number of physical properties and chemical reactions.

Alternatively, the organic portion of an organometallic compound may be viewed as a ligand that is bonded to the metal atom through carbon. This is the "inorganic viewpoint," which includes organic groups with a large variety of inorganic anions and molecules as species linked to the metal atom. In this approach it is the properties of the metal that are primary, and the organic groups merely cause some modifications. Such a viewpoint can also explain a substantial number of physical properties and chemical reactions.

Each viewpoint has both strengths and limitations. Probably the most important drawback of either viewpoint is the emphasis on one portion of the compound at the expense of the other (eg, *ORGANO*–metallic versus organo–*METALLIC*), thereby giving a distorted perspective. Many research projects do emphasize the organic or the metal portion of the compound, but such an approach causes difficulty in discussing the field as a whole. Therefore, this book will use a third approach, the "tie that binds" viewpoint, in which the metal–carbon bond serves as the focal point for classification and exposition.

The very definition of an "organometallic" compound is neither uniform nor completely consistent—a point discussed by Ries.[2] He concluded:

> Frontiers, at a time when it has become possible to graft almost any organic fragment onto many metals (and conversely to introduce metal atoms into so many organic molecules) have become more elusive than ever . . .

In its simplest form, an organometallic compound might be defined as "any compound containing at least one metal–carbon bond." Unfortunately, questions arise when this definition is studied closely. Does the term "metal" include elements such as silicon, boron, phosphorus or arsenic, all of which have extensive organo derivatives but which are not considered metals? Does a species that has *any* carbon atom bonded to a metal fall into the category of "organometallic?" Section 29 of *Chemical Abstracts* ("Organometallic and Organometalloidal Compounds") defines its subject material as any "compounds that contain one or more carbon–metal or carbon–metalloid covalent (sigma and pi) bonds."

As far as this book is concerned, the "metal" portion of "organometallic" will include any element whose electronegativity is less than that of carbon. Concepts of electronegativity and metal character are developed and discussed in most standard texts on inorganic chemistry. This definition will include the great majority of elements that chemists consider true metals, along with certain elements (boron, silicon, germanium, phosphorus, arsenic and tellurium) that are not considered as metals. These elements are often termed "metalloids," and the term "organometalloid" has been used with increasing frequency to describe organo derivatives of these elements. Should organopolonium or organoastatine compounds ever be reported, they would also fall into this category, as would any organoxenon compounds. The organo derivatives of the most electronegative elements (fluorine, oxygen, chlorine, nitrogen and bromine) are considered as part of organic chemistry, although the terms "organofluorine" and "organochlorine" are receiving increasingly common usage. Sulfur, selenium and iodine all have very extensive organic chemistry, and are very close to carbon in their electronegativities. Generally, if these elements are in negative oxidation states, their organo derivatives are considered as organic compounds. However, organo derivatives containing these elements in positive oxidation states (eg, $(CH_3)_3SeCl$, $C_6H_5I(O_2CCH_3)_2$) are virtually organometalloidal compounds, and will be discussed as such. Recently the first organo derivative of trivalent bromine was isolated;[3] it would also fall into this category. Even the hydrocarbons themselves might, by stretching the definition, be considered "organometalloids," as hydrogen is less electronegative than carbon. In fact, certain properties of some organometals are best considered by analogy or extension of corresponding properties of hydrocarbons.

The carbon atoms that comprise the other half of metal–carbon bonds come in a variety of environments. The "classical" organo groups—the simple alkyl and aryl derivatives—are those formed by replacement of one or more carbon-hydrogen bonds by metal(loid)–carbon bonds, as are found in C_6H_5Na, $(CH_3)_2Zn$, $(C_2H_5)_4Pb$, $(C_6H_5CH_2)_4Si$, etc. In all these compounds, the carbon atom remains tetravalent. Another enormous class of organometallic compounds are those in which the organic group is bonded to the metal through the electrons in carbon–carbon pi-bonds. The first and best-known example in this category is ferrocene, $(C_5H_5)_2Fe$. The carbon atoms in these compounds are also tetravalent. Presumably the growing number of intercalation compounds, in which metal atoms or compounds insert between the carbon layers in graphite and interact with the electron density there, would also fall into this category.

Many metal derivatives of nominally divalent carbon atoms (carbon monoxide, isocyanides, carbenes) have also been reported; these also are considered organometallic compounds. Binary metal carbides have not been treated consistently: those of representative metals, such as calcium acetylide, CaC_2, usually are considered organometallic, while those of transition metals, especially the "nonstoichiometric" carbides, are not. Metal cyanides and metal fulminates are usually not considered organometallic compounds, even though they undoubtedly contain metal–carbon bonds.

Finally, a fair number of esters of the oxy acids of elements of Groups

IV through VI have been lumped under the category of organometalloids; for example, tetraethylpyrophosphate, $(C_2H_5O)_4P_2O_3$, is one of a large number of compounds referred to as "organophosphorus" compounds by many. Compounds of this type, which contain no metal(loid)–carbon linkage, will not be considered in this book. As Ries points out,[2] in a dynamic and growing area such as organometallic chemistry, which impinges on so many other areas of chemistry, consistency is not always to be expected, or even necessarily desirable. Then, too, usage changes. Although compounds such as $Ni(CO)_4$ have been known for nearly a century, they have been considered as organometallic compounds only in the last third of that period. Doubtless there will be additional changes in years to come.

1.2 NOMENCLATURE OF ORGANOMETALLIC COMPOUNDS

The terms "organometallic" and "organometalloidal" have already been defined. A comparable earlier term now rarely seen is "metalloorganic." In 1855 *Chemisches Zentralblatt* introduced the term "organische metallverbindungen" into its index and nine years later added the further term, "metallorganische verbindungen." Occasionally one encounters the term "metal–organic." This is a more general term that incorporates all linkages between metal atoms and organic groups, regardless of what atom of the organic portion may be involved, and would include such compounds as sodium acetate, $K^{+-}NHC_6H_5$, $Hg(SCH_3)_2$, etc. The prefix "organo–", when combined with the name of a metal or metalloid, designates those compounds containing a direct linkage between a carbon atom and the element involved, eg, organomagnesium, organo– silicon, organomercury, etc. Occasionally this might be used for all elements, and the term "organoelement" appears in the Russian chemical literature.

Most common organometallic compounds may be named simply by combining the name(s) of the organic radical(s), using Greek prefixes, if necessary, with the name of the metal. These should be spelled as a single word, although hyphenated spellings are encountered in the earlier literature. Once in a while it may be necessary to designate the oxidation state of the metal; this is done by using a Roman numeral in parentheses, eg, dimethylplatinum(II). Generic terms are handled in the same way (eg, alkyllead, arylchromium). If inorganic groups are also present as ligands on the metal atom, these follow the name of the metal and are written as separate words (eg, $(C_6H_5)_3BiCl_2$, triphenylbismuth dichloride). Substituents on the organic group are usually included in the name of the organo substituent (eg, $ClCH_2CH_2AlCl_2$, 2-chloroethylaluminum dichloride). Once in a while, especially in the earlier literature, the metal may be named as a substituent on the organic group. This is most commonly encountered for organomercury compounds (eg, 4-ClHgC$_6$H$_4$CO$_2$H, p-chloromercuribenzoic acid). Other examples of these rules are listed in Table 1.1.

Organometalloidal compounds are named by analogy with methane derivatives, in that the root is not the name of the metalloid but of the corresponding hydride (eg, BH$_3$, borane; SiH$_4$, silane; GeH$_4$, germane; PH$_3$, phosphine; AsH$_3$, arsine). Thus (CH$_3$)$_4$Si is tetramethylsilane rather than tetramethylsilicon. Since phosphorus and arsenic have organo derivatives in their pentavalent oxidation state, the two hypothetical hydrides PH$_5$, phosphorane, and AsH$_5$, arsorane, have been used in recent years as root names (eg, (C$_6$H$_5$)$_5$P, pentaphenylphosphorane, although pentaphenylphosphorus is also used). Polynuclear organometalloids are named as the substituted derivatives of the corresponding polynuclear hydrides (eg, Si$_2$H$_6$, disilane; Ge$_3$H$_8$, trigermane; As$_4$H$_6$, tetraarsine, etc.). If the compounds are cyclic, the prefix "cyclo" appears before the name of the hydride (eg, (CH$_3$)$_{10}$Si$_5$, decamethylcyclopentasilane). Occasionally a metal or metalloid may be part of a ring of carbon atoms; in this situation, the metal or metalloid is named as an infix (eg, 1,1-dimethyl-1-silacyclopentane).

Binary compounds between metals and carbon monoxide have the metal named first, followed by the term "carbonyl"; either or both may use numerical prefixes as needed (eg, Fe$_3$(CO)$_{12}$, triiron dodecacarbonyl). Any inorganic groups on the metal atoms follow the carbonyl moiety (eg, HCo(CO)$_4$, cobalt tetracarbonyl hydride), but alkyl or aryl groups are named according to the rule already given (eg, CH$_3$Mn(CO)$_5$, methylmanganese pentacarbonyl). Table 1.2 lists some additional examples. The rules for the extensive series of compounds between metals and unsaturated organic molecules will be discussed in later chapters.

Table 1.1. Molecular Formulas and Names of Some Organometallic Compounds

Formula	Name
C$_3$H$_7$Li	n-propyllithium
(C$_6$H$_5$)$_2$Hg	diphenylmercury
(C$_2$H$_5$)$_3$Ga	triethylgallium
(C$_6$H$_5$CH$_2$)$_4$Pb	tetrabenzyllead
(C$_6$H$_5$)$_5$Bi	pentaphenylbismuth
(CH$_3$)$_6$W	hexamethyltungsten
(C$_2$H$_5$)$_6$Sn$_2$	hexaethylditin
(CH$_3$Sb)$_5$	pentamethylcyclopentaantimony
C$_2$H$_5$MgBr	ethylmagnesium bromide
(C$_6$H$_5$)$_2$TlCN	diphenylthallium cyanide
(C$_2$H$_5$)$_2$PbCl$_2$	dibutyllead dichloride
(CH$_3$)$_3$PtI	trimethylplatinum iodide
Fe(CO)$_5$	iron pentacarbonyl
Co$_2$(CO)$_8$	dicobalt octacarbonyl
Ru$_3$(CO)$_{12}$	triruthenium dodecacarbonyl

Table 1.2. Molecular Formulas and Names of Some Organometalloidal Compounds

Formula	Name
$(CH_3)_3B$	trimethylborane
$(C_2H_5)_4Ge$	tetraethylgermane
$(C_6H_5)_3As$	triphenylarsine
$(C_3H_7)_3SiCl$	tri-n-propylchlorosilane
$(C_6H_5)_3PBr_2$	triphenylphosphine dibromide; dibromotriphenylphosphorane
$(CH_3)_2TeI_2$	dimethyltellurium diiodide
$(CH_3)_4B_2H_2$	tetramethyldiborane
$(C_6H_{13})_8Si_3$	octahexyltrisilane
$(C_6H_5As)_4$	tetraphenylcyclotetraarsine
$(CH_3)_3SiGe(C_2H_5)_3$	trimethylsilyltriethylgermane; triethylgermyltrimethylsilane
$C_6H_5B(OH)_2$	phenylboronic acid; benzeneboronic acid
$CH_3As(:O)(OH)_2$	methylarsonic acid; methanearsonic acid
$(C_2H_5)_2P(:O)OH$	diphenylphosphinic acid
$Cs^+ (C_6H_5)_4B^-$	cesium tetraphenylborate
$(CH_3)_4As^+ I^-$	tetramethylarsonium iodide

Common or trivial names are relatively infrequent in organometallic chemistry, but those that do occur are used quite frequently. For the most part, they are historical in origin. "Cacodyl," for example, was originally coined in the belief that the compound now known to be $(CH_3)_4As_2$(tetramethyldiarsine) was a stable radical. While names such as cacodyl chloride (for $(CH_3)_2AsCl$) occurred in the early chemical literature, currently this root survives only in the name "cacodylic acid" for the compound $(CH_3)_2AsO_2H$, also known as dimethylarsinic acid. The name "Zeise's salt" is given to the compound $K^+ C_2H_4PtCl_3^-$, in honor of the man who first prepared it; the related compound $(C_2H_4PtCl_2)_2$ is known as "Zeise's dimer." The term "Grignard reagent" refers to a group of compounds of general formula RMgX; occasionally a derived term such as "methyl-Grignard" (for CH_3MgBr) is used informally. When the compound $(C_5H_5)_2Fe$ was first reported, it received the common name "ferrocene." The enormous growth in related compounds led to a variety of analogous names for related species, eg, cobaltocene, nickelocene, etc. For compounds such as $(C_5H_5)_2TiCl_2$, names such as "titanocene dichloride" are often used. "Metallocene" is similarly used as a generic term.

Industrial or medicinal preparations containing organometal(loid)s often receive trade names. "Mercurochrome" and "merthiolate" are antiseptics containing organomercury compounds and are household names. "Salvarsan" was an organoarsenical—the first antibiotic—and was widely used in the treatment of syphilis. Commercial names for biocidal organotin compounds are given in Chapter 12. Lewisite($ClCH=CHAsCl_2$) was used as a poison gas in World War I; Sarin and Soman are nerve gases containing phosphorus–carbon linkages.

1.3 HISTORICAL DEVELOPMENT OF ORGANOMETALLIC CHEMISTRY

A thorough presentation on the chronology of organometallic chemistry and its relationships to other areas of chemistry has been published elsewhere;[4] only a brief summary will be presented here. For convenience in presenting this summary, the three dates 1850, 1900, and 1950 will be used as reference points.

Prior to 1850

The first compound identifiable as an organometal was a methylarsenical, first prepared in solution by Cadet de Gassicourt in 1760 ("Cadet's fuming arsenical liquid"), and investigated in detail by Robert Bunsen in the period 1837–1843, at considerable risk to his health. The compound was so reactive that it was thought to be an isolated free radical and was given the name "cacodyl" because of its foul odor; later the stoichiometry was found to be $(CH_3)_4As_2$. About this same time the Danish pharmacist Zeise isolated and reported a platinum–organic compound, now known as Zeise's salt. After considerable controversy, its stoichiometry was established to be K^+ $C_2H_4PtCl_3^-$ the first isolated compound containing an unsaturated organic molecule bonded to a metal. In 1849 Edward Frankland, working in Bunsen's laboratory, prepared and isolated diethylzinc, $(C_2H_5)_2Zn$. This volatile, inflammable liquid was the first organometal to be used as a reagent, and opened up a new phase in the development of organometallic chemistry.

1850 to 1899

The availability and reactivity of diethyl- and dimethylzinc enabled the synthesis of alkyl derivatives of many other metals and metalloids. Within a very few years, enough such compounds had been prepared and isolated that Frankland was able to use their properties when he formulated his theory of valence. Development continued to the point where in 1870 Mendeleyev could use the known properties of tetraethylsilane and tetraethyltin to predict the properties of the corresponding derivative of "ekasilicon" (now germanium) with considerable accuracy, as part of his formulation of the Periodic Law. Numerous additional alkyl- and arylmetal compounds were prepared. In 1890 Ludwig Mond, investigating the unusually rapid corrosion of nickel valves used in apparatus for the industrially important Solvay process, reported nickel tetracarbonyl, $Ni(CO)_4$, the first of the metal carbonyls to be isolated. Early investigators reported toxic effects of organometals, including scattered fatalities. Organomercury compounds first appeared in the pharmacopeia as useful in the treatment of syphilis. Goiso, investigating the nature of so-called "arsenic rooms"

(rooms in which people came down with arsenic poisoning from no visible cause), found that a malodorous gas emanated from mold-infested wallpaper; he formulated this gas as an organoarsenic compound, and it became known as "Gosio-gas."

1900 to 1950

The new century opened with a crucial discovery that would have an enormous effect on both organometallic chemistry and organic chemistry generally. Victor Grignard reported the preparation and synthetic uses of solutions of alkylmagnesium halides in ethers. These solutions, collectively termed Grignard reagents, subsequently became *the* organic reagent. The Grignard reagent quickly displaced dialkylzincs in preparative reactions because it was easier to prepare, easier to handle and considerably more effective. Its availability accelerated the development of organometallic compounds as reagents in organic synthesis, a research area that remains active to this day. While many additional organometallic compounds were prepared and reported for the first time during this period, emphasis shifted from synthesis to characterization, especially with respect to structure and bonding. Over this period also, laboratory techniques showed increasing refinement and flexibility, enabling the study of compounds and reaction intermediates (such as Paneth's generation of organometals from the reaction of radicals with metal coatings) impossible under earlier conditions.

Applications of organometallic compounds outside the laboratory also appeared around this time: tetraethyllead as a gasoline additive; silicones in various applications; metal carbonyls as reaction catalysts; alkylaluminum compounds as intermediates in polymer synthesis. The extensive work by Ehrlich and co-workers on organoarsenic compounds led to the preparation and use of Salvarsan (4-hydroxy-3-aminophenylarsenobenzene polymer) for the treatment of syphilis and related afflictions, and also laid the foundations for the subsequent development of chemotherapy. Organoarsenicals were also used as poison gases in World War I. During the 30's and 40's, Challenger characterized "Gosio-Gas," showing that it was trimethylarsine, $(CH_3)_3As$; more importantly, he showed that this formed through direct biological action of molds on arsenic trioxide (used in earlier times as part of wallpaper coloring agents). This was the first reported example of the formation of a carbon-metal(loid) bond through the medium of a living organism.

1950 and Thereafter

The independent preparation of ferrocene, $(C_5H_5)_2Fe$, by two different research groups in 1951-1952 has had an effect on the development of organometallic chemistry at least as great as the Grignard reagent. This compound was the precursor of an enormous and continually growing number of transition metal organo derivatives. Equally importantly, it provided the impetus to tie together many previously disparate compounds whose structures and/or bonding were unclear through the application of bonding theory to organometallic compounds. Investigations in this area were abetted by, and in turn contributed to, the rapid development in many instrumental techniques (vibrational spectroscopy, resonance spectroscopy, mass spectrometry, x-ray crystallography, to name a few). Very shortly thereafter, these in turn would be aided by the rapid growth of computer technology and applications. Transition metal chemistry expanded enormously with the development of metallocenes, metal carbonyl compounds and, more recently, cluster compounds.

The impetus provided by the discovery of ferrocene pervaded virtually every area of organometallic chemistry. Earlier industrial applications expanded and diversified, especially silicone polymers and more recently organometallic catalysts. Synthetic techniques have been developed and refined to a degree that would have seemed incredible before 1950, with the result that "unstable" or "nonexistent" compounds have been isolated and characterized. The proliferation of instrumental techniques has facilitated investigations into the structure and bonding of organo derivatives of main-group elements. New compounds are not only appearing at a rapid rate, they are being characterized with a thoroughness and sophistication impossible thirty or forty years ago.

A recent development whose full influence remains to be determined has been the role of organometallic compounds in biology and the environment. The tragic cases of poisoning by methylmercuric compounds in Japan ("Minamata Disease") and elsewhere has generated substantial research efforts on these compounds, especially after it was discovered that they could be formed by biological processes and were virtually ubiquitous in natural waters. Increasing use of organotin compounds, especially tri-n-butyltin species, as antifouling agents, with their resulting introduction into the environment, has stimulated research on the biological effects of these compounds. The development of increasingly sensitive analytical techniques has shown the existence of unexpected and unsuspected organometal(loid)s in nature and has made possible their investigation at much lower concentrations than ever before. Organo derivatives of the metalloids, especially phosphorus, have proven useful in chemotherapy and in biochemical investigations.

At this point in time it is virtually impossible to give any complete assessment of the present position of organometallic chemistry and its development since the discovery of ferrocene. Suffice it to say that the field has achieved an energy and diversification virtually unmatched by any other area of chemistry, and that the rate of change seems to be accelerating.

1.4 THE LITERATURE OF ORGANOMETALLIC CHEMISTRY

The extensive research activity involving organometallic compounds has generated an accumulation of papers, reviews, monographs, etc., that is both voluminous and widely dispersed. The extent of this development can be seen in the following illustration.

In 1937 Krause and von Grosse published their comprehensive book "Die Chemie der Metall-Organischen Verbindungen" ("The Chemistry of Metal-Organic Compounds"); it was a single volume some 900-plus pages long. Forty-five years later the collection "Comprehensive Organometallic Chemistry" appeared in print, consisting of eight volumes with 8800 pages, plus an index volume of 1570 pages!

Various compilations of organometallic compounds have appeared; the most recent of these are listed at the end of this chapter. Regular review series exist; the most enduring has been "Advances in Organometallic Chemistry," a hardcover collection, which first appeared in 1961 and is still publishing regularly. Monographs on the organo derivatives of individual metals have been written for quite a few different elements; these will be cited at appropriate places in this book.

Two journals currently publish exclusively in the area of organometallic chemistry: *Journal of Organometallic Chemistry* (Elsevier), which includes subject reviews and annual surveys; and *Organometallics* (American Chemical Society). A third such journal *Applied Organometallic Chemistry* (Longmans), will begin publication in 1987. Many papers in organometallic compounds appear in journals devoted to inorganic chemistry or organic chemistry. Less frequently, organometallic compounds are the subject of more specialized studies that are published in journals of other areas of chemistry, or even outside the area of chemistry itself, especially in the biological, medicinal and environmental sciences. Section 29 of *Chemical Abstracts* is devoted exclusively to organometallic compounds, and is the best place to follow trends in current research.

1.5 THE CLASSIFICATION OF ORGANOMETALLIC COMPOUNDS

Even the most cursory examination of the entire field of organometallic chemistry shows great diversity in compounds and their various properties. In order to discuss them with any degree of consistency or order, some method of classification needs to be adopted. As stated earlier, this book will use the metal–carbon bond as the characteristic property. On this basis the following classes of such bonds exist:

Ionic — The metal–carbon bond is ionic in nature, with a metal cation and a carbanion.

Electron-deficient — The metal–carbon linkage involves the sharing of an electron pair on a single carbon atom with two metal atoms.

Covalent — The metal(loid)–carbon linkage involves the sharing of two electrons between a carbon atom and a metal(loid) atom. The most common type of bond.

Synergistic — The metal–carbon linkage involves interaction of pi electrons of an olefin (or nonbonding electrons on a divalent carbon atom) with empty orbitals on metal atoms. This type of linkage is essentially confined to transition metals.

These bond types are not necessarily mutually exclusive. There are many compounds that can show two different types of metal–carbon linkages. Similarly, the same metal atom can change from one type of linkage to another. This classification cuts across the Periodic Table, but does emphasize similarities among compounds of the same bond type. The last two categories are extremely large, and subdivisions are used for convenience in discussion.

While any system of classification of this enormously varied discipline will have some drawbacks, this system has fewer than most and its own unique strengths.

REFERENCES

1. Kraus, C. A., *J. Chem. Educ.* **1929**, *6*, 1478.
2. Ries, J. G., *J. Organometal. Chem.*, **1985** *281*, 1.
3. Nguyen, T. T.; Martin, J. C., *J. Am. Chem. Soc.*, **1980** *102*, 7382.
4. Thayer, J. S., *Adv. Organomet. Chem.*, **1975** *13*,1.

BIBLIOGRAPHY

General Collections

Wilkinson, G.; Stone, F. G. A.; Abel, E. W., eds., "Comprehensive Organometallic Chemistry," Pergamon: Oxford, 1982, 9 volumes.
"Dictionary of Organometallic Compounds," Chapman & Hall: London, 1984, 3 volumes.
Kaufman, H. C., "Handbook of Organometallic Compounds," Van Nostrand: Princeton, 1961.

Review Series

Stone, F. G. A.; West, R., eds., "Advances in Organometallic Chemistry," Academic: New York, 1961ff, 25 volumes to date.
Becker, E. I.; Tsutsui, M., eds., "Organometallic Reactions," Wiley-Interscience: New York, 1970ff, 5 volumes to date.
Seyferth, D., ed., "Organometallic Chemistry Reviews," Elsevier: Amsterdam, 1977ff., 9 volumes to date.

Chapter 2

The Metal–Carbon Bond: Methods of Synthesis

2.1 GENERAL CONSIDERATIONS

Synthetic methods that include or utilize organometallic compounds number many hundreds. Many of these involve alteration or transfer of the organic portion, and fall into the province of synthetic organic chemistry. Others concern exchange of inorganic groups attached to the metal atom, or coordination of the metal itself to some other atom; such reactions are more appropriately considered the province of inorganic chemistry. For the purposes of this text, syntheses of concern will be those which are used to form a metal(loid)–carbon linkage. Such syntheses may be divided into two broad categories for purposes of discussion:

1. Reactions in which the metal/metalloid reactant exists in its elemental form
2. Reactions in which the metal/metalloid reactant exists as an already-formed chemical compound

Each type of synthesis has specific advantages and drawbacks. Syntheses in the first class take advantage of the greater reactivity of the metal itself, relative to the metal's compounds, and the fact that elements usually have a considerable quantity of "stored" energy, which can be released in reaction, thereby provides a considerable driving force. Metal compounds, on the other hand, are usually easier to handle, more available and less expensive. In planning the synthesis of a specific compound, the chemist takes into account a variety of factors, including availability of starting materials, reaction conditions (including the need for any special equipment), the reaction yield, the properties of the product (if known), the need for isolation (if the organometal product itself is to be used as a reagent, it would be treated somewhat differently than if it were to be isolated and characterized), etc. Theoretically any reaction in which an organometal(loid) forms as a product might be used for synthesis. In practice, however, many are not used for this purpose because they are deficient in some important respect, or perhaps some other reaction does the same thing better, faster or less expensively.

Following are the more common classes of organometal syntheses. Discussion is, of necessity, brief and generalized; readers interested in the preparation of some specific compound should check the chemical literature for the detailed conditions of its synthesis.

2.2 SYNTHESES USING ELEMENTAL METALS

2.2.1 The Direct Synthesis[1]

The Direct Synthesis is probably the most general and most important preparative method for organometallic compounds. It is widely used in laboratory preparations, and has been extended to industrial-scale syntheses. In the simplest form, a metal reacts with an organic halide:

$$M + n\,RX \longrightarrow R_nMX_n \qquad (2.1)$$

In some instances, the organometal halide may undergo disproportionation, and the actual reaction becomes:

$$2\,M + n\,RX \longrightarrow R_nM + MX_n \qquad (2.2)$$

Rates and yields of reactions vary widely, depending on:

a. *Metal* — The less electronegative the metal, the more readily the reaction proceeds and the milder the reaction conditions. Certain metals, most notably Mg and Al, form resistant oxide coatings on their surface that lower their reactivity substantially. Metalloids, which usually have strong covalent bonds between atoms, need more vigorous conditions for reaction.

b. *Halide* — Chlorides or bromides are the most commonly used. The carbon–fluorine bond generally requires too strenuous conditions for satisfactory reaction rates, while organic iodides tend to undergo coupling reactions to form carbon–carbon bonds. Alkyl groups having a hydrogen atom on the carbon adjacent to the carbon with the halogen may undergo dehydrohalogenation with active metals, giving an olefin as product. Aryl halides generally are less reactive than alkyl halides, but both groups show considerable variation depending on other substituents on carbon.

c. *Reaction Conditions* — These vary widely from system to system. In most cases, a solvent is used. This is best known for the Grignard Reagent, probably the most widely used example of the Direct Synthesis. Diethyl ether is commonly used as solvent. The strong solvation of the magnesium by the oxygen atom of the ether provides the driving force, enabling mild reaction conditions to be used and the reagent to be generated in a conveniently handled solution.

Two examples of the Direct Synthesis have important industrial applications. The first is the Direct Process Reaction, used to prepare the precursors to the silicones:

$$2 \ CH_3Cl \ + \ Si \ \xrightarrow[Cu]{300°} \ (CH_3)_2SiCl_2 \tag{2.3}$$

The Cu—Si alloy used must be prepared in a certain way, and the reaction conditions carefully controlled. Forty-one additional products can be formed,[1] but $(CH_3)_2SiCl_2$ can be obtained in yields up to 80%. Although this reaction is generally found only in industry, it can occasionally be used in research laboratories, as the following example shows:[2]

$$CH_2Cl_2 \ + \ Si \ \xrightarrow[Cu]{320°} \ Si_nC_nH_{4n+2} \qquad (n = 4\text{-}9) \tag{2.4}$$

Again, an assortment of products formed; the above is used only as a single example. The second industrial application involves the preparation of tetraethyllead as a gasoline additive:

$$4 \ C_2H_5Cl \ + \ Pb \ + \ 4 \ Na \ \xrightarrow{60°} \ (C_2H_5)_4Pb \ + \ 4 \ NaCl \tag{2.5}$$

In this preparation, a lead–sodium alloy is used, because lead, like most heavy metals, does not undergo the Direct Synthesis under ordinary conditions. This, however, seems to be due primarily to kinetic conditions, because finely divided lead particles react quite readily with alkyl halides. The formation and use of activated metals in the Direct Synthesis has become an increasingly important area of research.[3]

2.2.2 Metal–Hydrocarbon Reactions

Hydrogen has a lower electronegativity than carbon, suggesting that hydrocarbons might act as protonic acids. In point of fact, however, the great majority of hydrocarbons are such weak acids that only the strongest reactants can bring out their protonic properties. Metals can do this, as shown in the following reaction:

$$2 \ (C_6H_5)_3CH \ + \ 2 \ Na \ \longrightarrow \ 2 \ (C_6H_5)_3CNa \ + \ H_2 \tag{2.6}$$

This reaction can be used to prepare alkyl and aryl derivatives of the more active metals of Groups IA and IIA. The acidity of a carbon–hydrogen linkage depends very much on the stability of the carbanion formed (see Chapter 3 for discussion of this point). Some hydrocarbons can react with less active metals; one of the two initial reports of ferrocene used this approach:[4]

$$2 \ C_5H_6 \ + \ Fe \ \xrightarrow{300°} \ (C_5H_5)_2Fe \ + \ H_2 \tag{2.7}$$

Unsaturated hydrocarbons can react with metals to form compounds without the loss of hydrogen—a fact that may be used to isolate (or at least stabilize) highly reactive intermediates; the first isolated derivatives of cyclobutadiene, for example, were compounds such as $C_4H_4Fe(CO)_3$. A special case of this involves the formation of anion radicals at low temperatures, discussed in Chapter 3:

$$C_{10}H_8 + Na \xrightarrow[\text{THF}]{-65°} Na^+ \, C_{10}H_8^{\cdot -} \qquad (2.8)$$

This can be used to prepare compounds in certain special cases.

2.2.3 Metal–Carbon Monoxide Reactions

A few metal carbonyls may be prepared by the direct reaction of a metal with carbon monoxide. The most important of these is nickel tetracarbonyl:

$$Ni_{(s)} + 4 \, CO_{(g)} \xrightarrow{\Delta} Ni(CO)_{4(g)} \qquad (2.9)$$

This reaction is known as the Mond Process in industry (named after the discoverer of nickel tetracarbonyl). It can be used to separate nickel from cobalt (which does not react with carbon monoxide under these conditions); the nickel tetracarbonyl is decomposed by heating at higher temperatures. This synthetic method might also be applicable for other divalent carbon compounds (eg, CS, RNC, etc.).

2.2.4 Metal Vapors

The surface area of a metal (or a metal compound) increases as it becomes more finely divided, resulting in enhanced chemical reactivity. Carried to its logical extreme, such subdivision would result in the breaking of all metal–metal interactions, leading to a vapor of single, isolated metal atoms. Such a vapor should be highly reactive and undergo reactions unknown for the corresponding bulk solid. This approach has received considerable attention in recent years.[5,6] Metal vapors may be used to prepare highly unusual and/or unstable organometals inaccessible by more conventional approaches, and virtually any organic molecule might serve as substrate. For the most part, this approach has been used to prepare very small quantities of product, and frequently the products are too reactive for ready isolation. Following are three examples of such preparations:

$$Fe_{(g)} + C_3H_6 + 3 \, PF_3 \longrightarrow C_3H_6Fe(PF_3)_3 \qquad (2.10)$$

$$Ni_{(g)} + 4 \, CS \longrightarrow Ni(CS)_4 \qquad (2.11)$$

$$Pt_{(g)} + C_6F_5Br \longrightarrow C_6F_5PtBr \qquad (2.12)$$

2.3 SYNTHESES INVOLVING METAL COMPOUNDS

2.3.1 General Considerations

Syntheses using metal compounds as starting materials are extremely common, particularly for the more electronegative or less reactive elements. For purposes of discussion, the synthetic reactions may be classified as:

Exchange — Interchange of one atom or group of atoms with another, resulting in the formation of metal–carbon bonds.

Addition — Conversion of a nonbonding pair of electrons, or a pi-bonding pair, into sigma-bonding electrons, resulting in the formation of metal–carbon bonds.

Most common syntheses will fall into these two categories. Occasionally an Elimination reaction (the opposite of Addition reaction) may be used to prepare a specific organometal, and there may be exceptional synthetic reactions that do not fall into the preceding categories.

As with the previous group of reactions, these syntheses cover a wide range of reaction conditions, and only the most general discussion can be presented here. Conditions for the preparation of a particular compound should be determined from the chemical literature.

2.3.2 Exchange Reactions

2.3.2.1 Carbon–Halogen Exchange

This synthetic method is probably the most widely used to prepare organometal(loid)s. In its most general form, it may be represented by the following equation:

$$R-m + m'-X \longrightarrow R-m' + m-X \qquad (2.13)$$

The lower-case m's indicate that there may be other groups bonded to the metal in question. In this preparation, the metal m must be more reactive and less electronegative than m'; and is usually an element of Groups IA or IIA. The Grignard reagent is the first choice for such preparations;[7] alkyllithium or alkylsodium compounds would be used for greater reactivity, while alkylaluminum or alkylzinc compounds would be used if lower reactivity is needed.

$$PCl_3 + 3\ CH_3MgBr \xrightarrow{\text{ether}} P(CH_3)_3 + 3\ MgBrCl \qquad (2.14)$$

$$ZrCl_4 + 4\ (CH_3)_3CLi \xrightarrow[-78°]{\text{ether}} Zr[(CH_3)_3C]_4 + 4\ LiCl \qquad (2.15)$$

$$HgBr_2 + (C_2H_5)_2Zn \longrightarrow (C_2H_5)_2Hg + ZnBr_2 \qquad (2.16)$$

In situations where the reactant m' has more than one halogen, a mixture of products is possible. Complete substitution requires an excess of the reagent R—m; otherwise, only partial substitution may result. If the reacting organometal is not reactive enough, partial substitution may occur regardless of the quantity of reagent present, as in the following example:

$$SiCl_4 + 3\ (CH_3)_2CHMgBr \longrightarrow (CH_3)_2CH_3SiCl + 3\ MgBrCl \quad (2.17)$$

Isopropyllithium would be needed to replace the fourth chlorine. The factors determining the degree of substitution are:

Concentration — The higher the proportion of reagent to metal halide, the greater the degree of substitution.

Reactivity of Reagent — The reactivity of alkylmetals varies in the order K Na Li Mg Al Zn Cd. The more active the reagent, the easier it is to get greater substitution.

Organic Group — Alkyl groups generally substitute more readily than aryl groups. Among the alkyls, primary alkyls substitute more readily than secondary, which in turn substitute more readily than tertiary.

Halogen — Chlorides and bromides are most commonly used, with the bromides being more easily exchanged. Iodides are rarely used for this purpose, while fluorides, for the most part, are too unreactive to undergo ready substitution.

The two metals m and m' need not necessarily be different. There are many examples of exchange involving the same element, as in the following:

$$(CH_3)_4Sn + SnCl_4 \longrightarrow (CH_3)_3SnCl + CH_3SnCl_3 \quad (2.18)$$

In such cases, a mixture of products is obtained. The reaction conditions may be controlled to obtain maximum yields of the desired compound.

2.3.2.2 Metal–Halogen Exchange

This method of synthesis is more limited than the preceding and has been studied primarily for lithium, although almost certainly it would occur for other active metals as well. This exchange is written as the following:

$$RCl + R'Li \longrightarrow RLi + LiCl \quad (2.19)$$

The lithium atom bonds to the carbon atom that is less basic, for reasons that are discussed in Chapter 4. Chlorides are used in this reaction, since alkyl bromides often give coupling reactions to form carbon–carbon bonds.

2.3.2.3 Metal–Hydrogen Exchange

This reaction, frequently termed *metalation*, obeys the principle that an acid displaces a weaker acid from the latter's salts. Since hydrocarbons may be considered as extremely weak acids, and since the straight-chain alkanes are the weakest of these (see Chapter 4), lithium or sodium derivatives of the straight-chain alkanes may be used to prepare corresponding derivatives of other hydrocarbons:

$$\text{n-}C_4H_9Li + C_6H_6 \xrightarrow{\text{solvent}} C_6H_5Li + C_4H_{10} \qquad (2.20)$$

Since it is the salt of butane (one of the weakest hydrocarbon acids) and is readily soluble in hydrocarbon solvents, n-butyllithium is the most commonly used reagent for this reaction. Solvents appropriate for such preparations include ethers, especially those having two or more oxygen atoms.

A somewhat different example of metal–hydrogen exchange is the *mercuration* reaction between mercuric acetate and aromatic hydrocarbons:

$$C_6H_6 + Hg(OAc)_2 \longrightarrow C_6H_5HgOAc + HOAc \qquad (2.21)$$

This reaction also occurs for thallium(III) acetate and lead tetraacetate.

In general, carbon–hydrogen bonds are rather unreactive towards substitution, but the reactivity varies enormously depending on what substituents happen to be present. Hydrogens of hydrocarbons bonded to transition metals are generally more reactive chemically.[8] Studies on carbon–hydrogen bond activation currently represent an active area of contemporary organometallic research.

2.3.2.4 Metal–Metal Exchange

This reaction is one example of a *displacement* reaction. An organometal, in which the metal is relatively unreactive, is treated with a more active metal. Alkylmercurials are generally used as reactants in this synthesis, due to their low reactivity:

$$(C_2H_5)_2Hg + Zn \longrightarrow (C_2H_5)_2Zn + Hg \qquad (2.22)$$

This reaction is performed in such a way as to drive it to completion and allow the product to be separated.

2.3.2.5 Coupling Reaction

In its most general form, the coupling reaction might be written:

$$A\text{—}m + B\text{—}X \longrightarrow A\text{—}B + mX \qquad (2.23)$$

The reactions discussed in Section 2.3.2.1 might be considered as special examples of this. Coupling examples are very numerous and take advantage of the fact that mX is usually an ionic salt (eg, NaCl), whose exothermic heat of formation provides a strong driving force. Reactions such as the following represent the use of coupling reactions to form metal–carbon bonds:

$$(C_6H_5)_3GeNa + C_2H_5Cl \longrightarrow (C_6H_5)_3GeC_2H_5 + NaCl \qquad (2.24)$$

This reaction is used primarily for systems where the organic group, due to its structure and/or substituents, would not easily form a metal–carbon bond by the more common approaches previously mentioned.

Coupling reactions are actually more important in organometallic chemistry than the preceding might indicate. They are commonly used to form organometals having metal–metal linkages. Many examples of this use will appear in subsequent chapters.

2.3.3 Addition Reactions

2.3.3.1 Hydrometalation

This reaction involves the breaking of a carbon–carbon pi-bond and a metal(loid)–hydrogen bond, followed by formation of metal(loid)–carbon and carbon–hydrogen bonds:

$$m{-}H + H_2C{=}CH_2 \longrightarrow m{-}C_2H_5 \qquad (2.25)$$

The most thoroughly investigated example of this is hydroboration. Addition of diboranes to olefins, forming organoboranes, serves as the first step for a wide variety of organic syntheses.[9,10] Corresponding addition reactions are known for the hydrides of many elements, although reaction conditions vary extremely widely. For example, the temperatures needed for addition of the series $(n{-}C_4H_9)_3MH$ to olefins vary in the following manner:

M:	Si	Ge	Sn	Pb
Temperature:	300° (catalyst)	120°	90°	0°

Hydrometalation plays an important part in a variety of industrial processes involving the use of transition metal compounds as catalysts. One step in the Oxo (see Chapter 8.5.1) reaction involves the following reaction:

$$HCo(CO)_4 + CH_2{=}CH_2 \longrightarrow C_2H_5Co(CO)_4 \qquad (2.26)$$

Since hydrogen has such great use in many industrial processes and since metal catalysts are absolutely essential as catalysts for many of these reactions of hydrogen, the role and scope of hydrometalation seems destined to increase with passing time.

2.3.3.2 Organometalation

Metal–carbon bonds of the most active metals may react with the pi-bond of olefins in a manner analogous to hydrometalation:

$$m{-}R + H_2C{=}CH_2 \longrightarrow m{-}CH_2CH_2R \qquad (2.27)$$

This is best known and most studied for organoaluminum compounds, and serves as a method of generating long-chain hydrocarbon polymers:

$$(C_2H_5)_3Al + CH_2{=}CH_2 \longrightarrow$$
$$(C_2H_5)_2AlC_4H_9 \xrightarrow{\text{repeat}} Al(C_nH_{2n+1})_3 \qquad (2.28)$$

Organometalation also occurs for other metals and is part of the reason these metals are used as catalysts in polymer formation.

2.3.3.3 Oxymetalation

This reaction, which involves the addition of a metal–oxygen bond to an olefin, is limited in scope, being best known for mercury:

$$Hg(OAc)_2 + H_2C{=}CH_2 \longrightarrow AcOCH_2CH_2HgOAc \qquad (2.29)$$

Thallium triacetate and lead tetraacetate also give this reaction. They are frequently used in organic synthesis, since the initial products decompose readily to give vicinal diacetates.

2.3.3.4 Oxidative Addition

This reaction requires that the metal reactant have two stable oxidation states separated by two units. In effect, oxidative addition converts a nonbonding electron pair into a bonding pair. This reaction occurs for metals of Groups IIIA through VIA in their lower oxidation states, for the Group VIII metals,[11] and probably for other metals as well. Chemically, this reaction is virtually the same as the Direct Synthesis (Section 2.2.1), with the difference that the metals are already partially oxidized. Alkyl halides are generally used:

$$GeI_2 + CH_3I \longrightarrow CH_3GeI_3 \qquad (2.30)$$

$$(CH_3)_3Sb + C_2H_5Br \longrightarrow (CH_3)_3SbC_2H_5{}^+ \ Br^- \qquad (2.31)$$

$$CH_3PtClL_2 + CH_3Cl \longrightarrow (CH_3)_2PtCl_2L_2 \qquad (2.32)$$
$$(L = donor\ ligand)$$

This synthetic method is widely used, especially for the metalloids.

2.3.3.5 Insertion Reactions

Insertion reactions might be considered a special case of oxidative addition, in that a divalent carbon atom reacts with a metal–carbon bond:

$$C{:} + m{-}CH_3 \longrightarrow m{-}C{-}CH_3 \qquad (2.33)$$

The name "insertion" comes from an alternative way of viewing this reaction—namely, that the divalent carbon species (carbon monoxide, carbenes, isocyanides, etc.) insert into the metal–carbon bonds. While this reaction is best known for transition-metal compounds, it is not confined to them:

$$CH_3Mn(CO)_5 + CO \longrightarrow CH_3\overset{\overset{\displaystyle O}{\|}}{C}Mn(CO)_5 \qquad (2.34)$$

$$(C_2H_5)_3SiCl + CH_2N_2 \longrightarrow (C_2H_5)_3SiCH_2Cl + N_2 \qquad (2.35)$$

As equation 2.35 shows, carbenes are sufficiently reactive to insert into metal(loid)–halogen or –chalcogen bonds also. Insertion reactions are also known for other molecules as well.

2.3.4 Elimination Reactions

While numerous elimination reactions are known for organometals, these are rarely used for synthetic purposes. Elimination reactions are the opposite of addition reactions, and also require that the metal(loid) have two stable oxidation states:

$$CH_3CCo(CO)_4 \longrightarrow CH_3Co(CO)_4 + CO \qquad (2.36)$$
$$\underset{O}{\overset{\|}{}}$$

$$(C_6H_5)_3BiCl_2 \xrightarrow{\Delta} (C_6H_5)_2BiCl + C_6H_5Cl \qquad (2.37)$$

2.3.5 Anionic Complex Formation

A carbanion has a nonbonding pair of electrons localized on a single carbon atom, and thereby qualifies as a Lewis base. Under appropriate conditions, these anions can react with neutral alkylmetals having available coordination sites to form anionic organometals:

$$(C_2H_5)_2Zn + C_2H_5Na \longrightarrow Na^+ (C_2H_5)_3Zn^- \qquad (2.38)$$

$$(C_6H_5)_3B + C_6H_5Na \longrightarrow Na^+ (C_6H_5)_4B^- \qquad (2.39)$$

$$[(CH_3)_3PtI]_4 + 12\ CH_3Li \longrightarrow 4\ Li_2Pt(CH_3)_6 + 4\ LiI \qquad (2.40)$$

Such complexes are becoming increasingly common among transition metals, and often such peralkyl species are more stable than the neutral metal alkyls (see Chapter 5). The metal acceptor compound need not have alkyl groups on it, merely one or more empty orbitals to form the bond to carbon.

2.3.6 Reductive Carbonylation

A number of metal carbonyls may be prepared by reducing a metal oxide with carbon monoxide:

$$Re_2O_7 + 17\ CO \xrightarrow{\Delta} Re_2(CO)_{10} + 7\ CO_2 \qquad (2.41)$$

Presumably the metal (or a low-valent metal oxide) forms as a reaction intermediate.

2.3.7 Electrochemical Synthesis

Electrochemistry has undergone a spectacular development in recent decades, and is an excellent way of generating high-energy species that would otherwise be difficult or impossible to obtain. It has not been used much for synthesis, but seems to be particularly promising for the preparation of compounds in unusual oxidation states and/or containing metal–metal bonds:[12]

$$2 \ (C_6H_5)_3SnCl + 2 \ e^- \longrightarrow (C_6H_5)_6Sn_2 + 2 \ Cl^- \qquad (2.42)$$

Electrochemistry might also be used to generate reactive intermediates without any intention of isolation; such intermediates frequently yield valuable information to investigators.

REFERENCES

1. Rochow, E. G., *J. Chem. Educ.*, **1966**, *43*, 58.
2. Fritz, G.; Woersching, A., *Z. Anorg. Allgem. Chem.*, **1984**, *512*, 131–163.
3. Rieke, R. D., *Accts. Chem. Res.*, **1977**, *10*, 301.
4. Miller, S. A.; Tebboth, J. A.; Tremaine, J. F., *J. Chem. Soc.*, **1952**, 632–633.
5. Turney, T. W.; Timms, P. L., *Adv. Organometal. Chem.*, **1977**, *15*, 53.
6. Dagani, R., *Chem. Eng. News* (April 8, 1985), 23.
7. Kharasch, M. S.; Reinmuth, O., "Grignard Reactions of Non-Metallic Substances," Prentice-Hall: Englewood Cliffs, New Jersey, 1954.
8. Webster, D. E., *Adv. Organomet. Chem.*, **1977**, *15*, 147.
9. Brown, H. C., *Adv. Organomet. Chem.*, **1973**, *11*, 1.
10. Cragg, G. M. L., "Organoboranes in Organic Synthesis," Marcel Dekker: New York, 1973.
11. Collman, J. P., *Adv. Organomet. Chem.*, **1969**, *7*, 53.
12. Dessy, R. E., et al., *J. Am. Chem. Soc.*, **1966**, *88*, 451–476, 5112–5134.

Chapter 3

Ionic Metal–Carbon Bonds

3.1 INTRODUCTION

Completely ionic metal–carbon bonds would consist of metal cations and carbanions as distinct and separated entities. Such compounds should, in principle, show characteristic properties of salts: conductivity when molten or in polar solvents; little or no volatility; solubility in polar liquids, etc. In practice, many ionic organometallics will show some degree of covalent character, and the dividing line between ionic organometallics and those having electron-deficient covalent bonds (described in Chapter 4) is not always clearly defined. Alkyl and aryl derivatives of the alkali metals and alkaline earth metals (except Li, Be and Mg) are included in this category, along with the cyclopentadienides of Mn and the lanthanide metals. Other individual compounds may also possess ionic metal–carbon bonds.

The chemical reactivity of ionic organometallic compounds depends primarily on the carbanion. Using Ziegler's classification, these may be divided into the following categories:

Localized Carbanions — These have the negative charge confined to a single carbon atom. Example: $H_3C:^-$

Delocalized Carbanions — These have the negative charge distributed over two or more carbon atoms. Example: $(C_6H_5)_3C:^-$

Radical Carbanions — These have one (occasionally two) electrons added to an aromatic molecule. Example: $C_{10}H_8^{\overline{\cdot}}$

Most carbanions have a single negative charge, but there are some (mostly in the category of radical carbanions) that have a double negative charge. Also, there are many organometallic ions, such as $(CH_3)_4Sb^+$, $(C_6H_5)_4B^-$, $Mn(CO)_5^+$, $(C_5H_5)_2Co^+$, etc. The charge belongs to the unit as a whole, and the metal–carbon bonds are either covalent or synergistic. Such species will be discussed elsewhere in this volume.

3.2 LOCALIZED CARBANIONS

These carbanions are formed from saturated hydrocarbons and certain unsaturated ones. The carbon–hydrogen linkage has a slight polarization, making the hydrogen somewhat protonic and, in principle, replaceable by active metals. Most organometals of this type have only a single hydrogen replaced (eg, C_2H_5Rb, $(C_6H_5)_2Sr$, etc.), but is is possible to replace more than one, as in calcium acetylide, CaC_2, or sodium methanide, Na_4C. These compounds, along with organothium compounds, have been reviewed in an article by Wardell.[1]

Compounds containing localized carbanions are prepared by the Direct Synthesis (Chapter 2.2.1):

$$RX + 2\ M \longrightarrow RM + MX \tag{3.1}$$

This preparation must be done under very carefully controlled conditions to minimize the Wurtz coupling reaction:

$$RX + RM \longrightarrow RR + MX \tag{3.2}$$

Preparations are usually done at low temperatures, with the metal in the form of a dispersion or an amalgam, and an ether-like solvent present.

Metal derivatives containing localized carbanions are probably the most reactive and least tractable organometals known. They are white solids, highly sensitive to air, involatile, and insoluble in solvents unless they react with them. X-ray studies indicate that methylpotassium and the corresponding compounds of rubidium and cesium have ionic structures of the nickel arsenide type,[1] while methylsodium may be tetrameric, like methyllithium.[1] Thermal stability varies somewhat, depending on the alkyl group; if a β-hydrogen is present, elimination can occur readily:

$$C_2H_5Na \longrightarrow C_2H_4 + NaH \tag{3.3}$$

Because of their high reactivity, these compounds are generally used for synthesis and rarely isolated. Some examples of syntheses involving these compounds are:

Coupling $CH_3K + R_3CBr \longrightarrow R_3CCH_3 + KBr$ (3.4)

Metalation $C_2H_5Cs + C_6H_5CH_3 \longrightarrow C_6H_5CH_2Cs + C_2H_6$ (3.5)

Transfer $CH_3Na + (CH_3)_2Zn \longrightarrow Na^+ (CH_3)_3Zn^-$ (3.6)

The reactivity tends to increase as the atomic weight of the metal increases, at least for the alkali metals.

3.3 DELOCALIZED CARBANIONS

If the carbanionic hydrocarbon contains pi-bonds that can interact with the negative charge, the net effect will be to spread the charge ("delocalize") over two or more carbon atoms, thereby lowering the charge on any single carbon atom. This in turn lowers the basicity and the reactivity. The greater the extent of delocalization, the greater is the reduction in the basicity of the carbanion, and the higher the acidity of the corresponding hydrocarbon. The metalation reaction (equation 3.5) has been used to determine the relative acidities of hydrocarbons, since metalation in effect is an acid–base reaction, and a stronger acid always displaces a weaker acid from the latter's salts. Much of the earlier work in this area is summarized in a monograph by Cram.[2] Not all hydrocarbons are weak acids. Cyclopentadiene (pK$_a$ = 15) is a stronger acid than water (pK$_a$ = 16). Certain cyclopentadienylmetal compounds are ionic, whereas the corresponding alkylmetal compounds are covalent; one example is manganocene, $(C_5H_5)_2Mn$. Like sodium cyclopentadienide, it can be used to prepare ferrocene:[3]

$$Mn(C_5H_5)_2 + FeCl_2 \longrightarrow Fe(C_5H_5)_2 + MnCl_2 \qquad (3.7)$$

Ferrocene and similar synergistically bonded metallocenes do not undergo this type of reaction, which serves as an indication of the presence of an ionic metal–cyclopentadiene linkage. Such linkages have been proposed for the cyclopentadienide derivatives of the lanthanide and actinide metals.[3]

The metalation reaction has been used as a technique for the separation of aromatic compounds having very similar boiling points.[4] For example, the equilibrium between the sodium derivative of p-xylene and m-xylene

$$NaCH_2 - C_6H_4 - CH_3 + CH_3 - C_6H_4 - CH_3 \longrightarrow$$

$$CH_3 - C_6H_4 - CH_2Na + p\text{-}CH_3C_6H_4CH_3 \qquad (3.8)$$

has an equilibrium constant of 9.2. The corresponding equilibrium between metalated ethylbenzene and p-xylene favors the formation of metalated p-xylene.

Methane, being quite an unreactive hydrocarbon, does not readily undergo metalation. However, an unusual form of the metalation reaction, involving the metal lutetium, has been reported.[5] The compound methylbis(h^5-pentamethylcyclopentadienyl)lutetium reacted with labeled methane in perdeuterocyclohexane:

$$(Me_5C_5)_2LuCH_3 + {}^{13}CH_4 \longrightarrow (Me_5C_5)_2Lu{}^{13}CH_3 + CH_4 \qquad (3.9)$$

This reaction is an example of numerous efforts to "activate" the rather sluggish carbon–hydrogen linkage, in order to find catalysts for the better utilization of hydrocarbons in chemical industry.[5]

3.4 RADICAL CARBANIONS

3.4.1 Anions of Aromatic Hydrocarbons

Compounds discussed in the preceding sections are *substitution* compounds, in which one (or more) hydrogens of a hydrocarbon have been replaced by a metal atom. Thus naphthalene might undergo metalation:

$$C_{10}H_8 + CH_3K \longrightarrow C_{10}H_7K + CH_4 \qquad (3.10)$$

However, at low temperatures in a suitable solvent, naphthalene can react with potassium to give quite a different reaction:

$$C_{10}H_8 + K \longrightarrow K^+ \, C_{10}H_8^{\,\tau} \qquad (3.11)$$

In this case, an electron has been transferred from the potassium atom to the naphthalene ring; no hydrogens are lost. The resulting product has an unpaired electron and is therefore a radical, as well as a negative charge; therefore, it is an example of an *anion radical*. The compound would be named potassium naphthalenide, whereas the product in equation 3.10 would be named α- (or β-) naphthylpotassium. The great majority of aromatic hydrocarbons can form such anion radicals, and they have received considerable attention.[6,7]

Anion radicals are typically prepared by treatment of a suitable hydrocarbon, such as anthracene, with an alkali metal (usually as an amalgam) at low temperature:

$$C_{14}H_{10} + Na/Hg \xrightarrow[-80°]{\text{solvent}} Na^+ \, C_{14}H_{10}^{\,\tau} \qquad (3.12)$$

Ethers containing a high oxygen–carbon ratio, such as diglyme, $CH_3OCH_2CH_2OCH_2CH_2OCH_3$, are used as solvents; these molecules bind to the sodium ion, separating it from the anion and helping to stabilize the product.

Aromatic hydrocarbons can form anion radicals because they have empty antibonding pi molecular orbitals of relatively low energy available for accepting the donated electron. The energy level diagram for the pi bonding and antibonding orbitals for naphthalene is shown in Figure 3.1. In naphthalene, the lowest energy transition ("A" in Figure 3.1) is a $\pi \rightarrow \pi^*$ transition, which absorbs in the ultraviolet region of the electromagnetic spectrum. Three bands are actually observed (with energies of 45,200, 35,000, and 32,100 cm^{-1} respectively), due to configuration interactions within the molecule.[8] By contrast, in the radical anion, the electron of lowest energy lies in an antibonding orbital, and can undergo a $\pi^* \rightarrow \pi^*$ transition. This requires considerably less energy; in sodium naphthalenide, the transition absorption occurs at 755 nm (corresponding to an energy of 13,200 cm^{-1}), which falls in the visible region of the electromagnetic spectrum. The compound, therefore, shows a green color. Other anion radicals are usually intensely colored also.

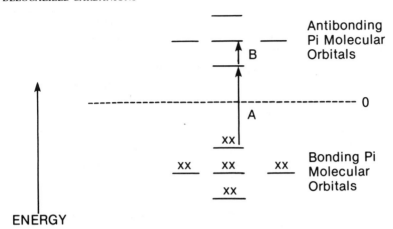

Figure 3.1. Energy levels for bonding and antibonding pi molecular orbitals in naphthalene.

The odd electron can be transferred from one hydrocarbon acceptor molecule to another. Workers have investigated this exchange. Sodium biphenylide in glyme is usually chosen as a reference point, since it readily transfers its electron to other aromatic hydrocarbons.[6] A partial list of the relative stabilities of anion radicals gives:

tetracene > anthracene > 1,4-diphenylbutadiene > pyrene > 1,2-diphenylethylene > phenanthrene > naphthalene > biphenyl

A pi-orbital (bonding or nonbonding) can hold two electrons, and it is therefore possible to make dianions (eg, 2 Na^+ $C_{10}H_8^=$). Depending on the hydrocarbon involved, these may be diradicals (the two electrons in two degenerate orbitals), or they may have their spins paired and occupy one orbital. The existence of dianions raises the possibility of the following equilibrium:

$$2 \ Ar^{\cdot} \ \rightleftharpoons \ Ar \ + \ Ar^= \qquad (3.13)$$

In general, this equilibrium lies far to the left. Radical anions are stable towards this type of disproportionation, unless there is some strong driving force that favors dianion formation. One exception is the anion radical of cyclooctatetraene, which undergoes ready disproportionation:

$$2 \ C_8H_8^{\cdot} \ \longrightarrow \ C_8H_8 \ + \ C_8H_8^= \qquad (3.14)$$

The aromatic 10-electron dianion is exceptionally stable, and this provides the driving force for the reaction.

Investigations into the properties of anion radicals have been greatly enhanced by the development of electron spin resonance (ESR) spectroscopy (also called electron paramagnetic resonance). This technique uses the spin of an unpaired electron as its experimental parameter. The absorption pattern varies with the extent of interaction of the electron spin with the spins of nearby nuclei. The resulting splitting patterns found in the ESR spectra of radical anions have provided much information about the interactions among the atoms in these molecules and about the energies of such interactions.

These anion radicals are stable only at low temperatures and decompose when warmed to room temperature with loss of hydrogen and formation of metalated products, which are also quite reactive. As shown in Figure 3.2, sodium naphthalenide reacts with water or trimethylchlorosilane to form both the 1,2- and 1,4-dihydronaphthalene derivatives.

3.4.2 Anions of Other Carbon-Containing Molecules

Aromatic hydrocarbons are not the only molecules that can form anion radicals. The essential requirement is that the acceptor molecule have a low-energy empty orbital available. Heterocyclic aromatic molecules form anion radicals. A commonly studied example is 2,2'-bipyridyl. The radical anion formed from this molecule can bond to metals, providing useful species for the study of metal–ligand interactions.

Certain organometallic compounds also form anion radicals. One early example of this was triphenylborane, which formed an anion radical isoelectronic with triphenylmethyl radical. This radical had some tendency to dimerize to form a boron-boron single bond:

$$2\ (C_6H_5)_3B^{\cdot -} \longrightarrow (C_6H_5)_6B_2^{=} \qquad (3.15)$$

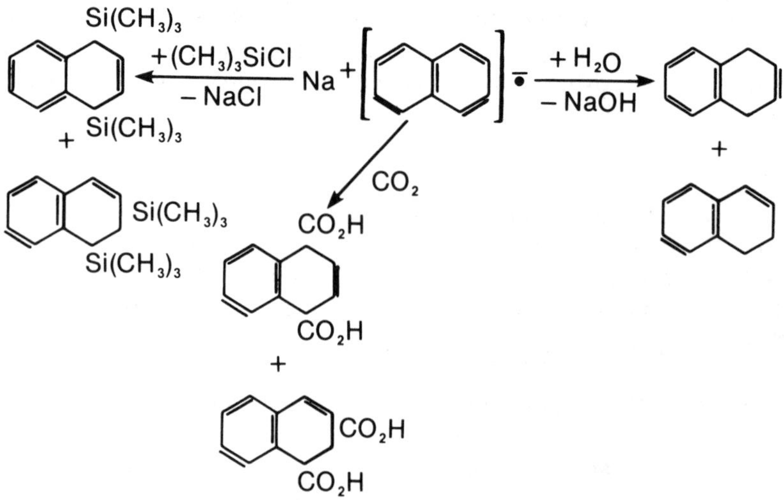

Figure 3.2. Reactions of sodium naphthalenide.

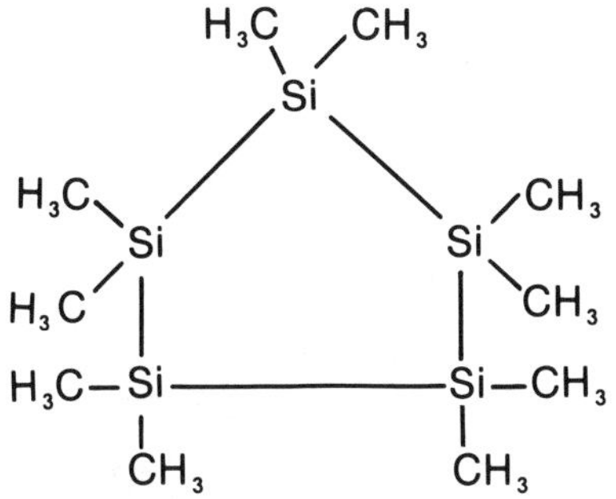

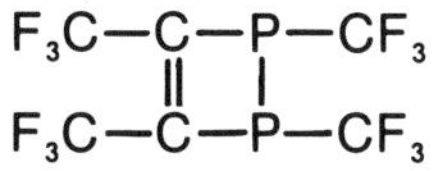

$$F_3C-C-P-CF_3$$
$$\|\quad\|$$
$$F_3C-C-P-CF_3$$

Figure 3.3 Two cyclic organometalloids capable of forming anion radicals.

Similar dimerization has been reported for radical anions derived from other Group III organo derivatives, such as the hexamethyldigallanide dianion, $(CH_3)_6Ga_2^=$.

The cyclic permethylcyclopolysilanes form anion radicals, and these have been studied using ESR spectroscopy.[9] It was found that $(CH_3)_{10}Si_5$ (shown in Figure 3.3) formed a radical anion in which the unpaired electron interacted equally with all thirty hydrogens present, suggesting a delocalization over the ring system.[9] The effect of this on the understanding of silicon–silicon bonding will be discussed in Chapter 7.

Another compound that forms an anion radical is the cyclic diphosphine shown in Figure 3.3. The ESR splitting patterns of this anion radical indicate that the unpaired electron resides primarily in the pi-antibonding orbital of the carbon–carbon double bond, and interacts only slightly with the phosphorus atoms.[10] This in turn suggests very little orbital interaction between phosphorus and carbon in this particular system (see Chapter 7 for a fuller discussion on this topic).

Still another organometallic system that forms anion radicals are the alkylcobalt carbonyls, $R_3CCo_3(CO)_9$. The ESR spectra of these radical anions indicate that the unpaired electron resides predominantly in an antibonding orbital in the Co_3 cluster.[11,12] Undoubtedly many more such radical anions will be prepared and investigated, with valuable additions to the knowledge of chemical bonding resulting.

REFERENCES

1. Wardell, J. L., in "Comprehensive Organometallic Chemistry," *1*, 1982, 43–120.
2. Cram, D. J., "Fundamentals of Carbanion Chemistry," Academic Press: New York, 1965, pp. 1–46.

 3. Maslowsky, E., *J. Chem. Educ.*, **1978,** *55,* 276.
 4. *Chem. Eng. News* (June 14, 1971), 30.
 5. Maugh, T. H., *Science,* **1983,** *220,* 1261.
 6. deBoer, E., *Adv. Organomet. Chem.,* **1965,** *2,* 115.
 7. Holy, N. L., *Chem. Rev.,* **1974,** *74,* 243.
 8. Jaffé, H. H.; Orchin, M., "Theory and Applications of Ultraviolet Spectroscopy," Wiley: New York, 1962, p. 287.
 9. West, R.; Carberry, E., *Science,* **1975,** *189,* 179.
10. Wallace, T. C.; West, R.; Cowley, A. H., *Inorg. Chem.,* **1974,** *13,* 182.
11. Peake, B. M.; Robinson, B. H.; Simpson, J.; Watson, D. J., *Inorg. Chem.,* **1977,** *16,* 405.
12. Peake, B. M.; Rieger, P. H.; Robinson, B. H.; Simpson, J., *Inorg. Chem.,* **1979,** *18,* 1000.

Chapter 4

"Electron-Deficient" Metal–Carbon Bonds

4.1 INTRODUCTION

The term "electron-deficient" bond is a historical misnomer that has persisted tenaciously. A variety of chemical systems have been reported in which three atoms are linked together by two electrons. These would be "deficient" only if the very common linkage of two atoms by two electrons is considered "normal." The technical term for these two systems would be "three-center, two-electron" and "two-center, two-electron bonds." Rundle was the first to publish a theoretical description of three-center bonds,[1] and is alleged to have remarked: "There is no such thing as an electron-deficient bond; there are merely theory-deficient chemists!" Nonetheless, the term "electron-deficient bond" has remained in common usage.

Various metals form bonds to carbon which fall under the classification of "three-center, two-electron" linkages. The majority of organo derivatives of lithium, beryllium and magnesium would be included, along with selected compounds of aluminum. Stable organometals of other metals having this type of bond are rare, but are commonly found as reaction intermediates. Three-center, two-electron bonds typically involve metals of fairly low electronegativity (but not quite low enough to give completely ionic metal–carbon bonds) with a valence shell that is less than half filled. Organometallic compounds having this type of bonding will show the following general properties:

1. *Extensive polymerization* — Especially in the solid state. The degree of polymerization will depend on the metal and the organic group primarily, with experimental conditions also affecting it.
2. *Complex formation with donor molecules* — Such formation is often quite exothermic, and plays an important part in the chemistry of these substances. Polymerization is usually less in donor solvents than in the solid state or in nondonor solvents.

3. *Chemical Reactivity* — These compounds have reactive metal–carbon bonds and are frequently used in organic syntheses. While less reactive than corresponding compounds of sodium, potassium or calcium, organometals in this category dissolve more readily in common solvents, making them considerably easier to handle.

4.2 STRUCTURE AND BONDING

Interaction of an atomic orbital of appropriate symmetry on each of three adjacent atoms leads to the formation of three molecular orbitals—almost always a bonding/nonbonding/antibonding combination, as shown in Figure 4.1. In "electron-deficient" organometals involving three atoms, the carbon is always the central atom and supplies the electrons.

The simplest compounds containing this type of bonding are the trialkylaluminum dimers, in which two groups form bridges between the two aluminum atoms. It should be noted that the electron density in such bridge bonds does *not* lie along the direct metal–carbon axes as in regular metal–carbon sigma bonds, but lies within these axes. This is a result of the geometries adopted to achieve maximum orbital overlap, and frequently lead to unusually small bond angles (eg, trimethylaluminum dimer has $\angle$ Al—C_{br}—Al = 74.7°, while polymeric dimethylberyllium has a bond angle of 66°).

The dialkyl compounds of beryllium and magnesium have the chain polymeric structure shown in Figure 4.1. All metal–carbon bonds are the three-center "electron-deficient" type. Such polymers are also known for various complexes containing different metals, such as $(Li_2ZnR_4)_x$. In these systems, while all metals have a coordination number of four, the bridging is not symmetric and the alkyl groups are more tightly bound to the more electronegative metal. Such compounds are intermediate between the completely covalent one-metal polymers, such as $[(CH_3)_2Be]_x$, and completely ionic systems, such as $K^+(CH_3)_4B^-$.

Organolithium compounds have the most complicated structures and have been the most thoroughly investigated.[2] In the solid state, these compounds usually exist as tetramers, with hexamers occurring less commonly and dimers being found in exceptional cases such as 7-norbornadienyllithium.[3] The framework of a tetramer (shown in Figure 4.1) has four lithium atoms at the corner of a tetrahedron, with the bridging carbon atoms of the alkyl group above the midpoint of each face. In effect, the structure is that of a tetracapped tetrahedron; each carbon atom interacts with three lithium atoms, and each lithium atom interacts with three carbon atoms. Organolithium hexamers, $(RLi)_6$, have six lithium atoms at the corners of a distorted octahedron, and six of the eight faces occupied by carbon atoms. The two unoccupied faces lie along a threefold axis of symmetry. Organolithium dimers have the same framework as trimethylaluminum dimer.

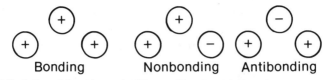

Bonding   Nonbonding Antibonding

Orbital Combinations in Three-Centered Bonds

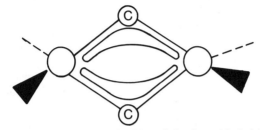

Electron Distribution in Metal-Carbon-Metal Bond

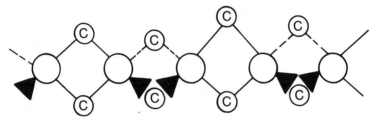

Chain Polymer Involving Three-Center Metal-Carbon Bonds

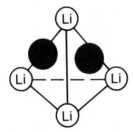

Alkyllithium Tetramer (two alkyl groups omitted for clarity)

Figure 4.1. Three-center two electron bonds and structures.

4.3 ALKYL EXCHANGE REACTIONS

Alkyl and aryl groups involved in three-centered metal–carbon bonding show considerable lability and undergo ready exchange, both through intramolecular and through intermolecular routes. Trimethylaluminum dimer was the first to be investigated in this regard, being studied by the then-novel technique of nuclear magnetic resonance (NMR) spectroscopy.[4] At room temperature, only a single peak was seen for this compound, even though its structure would predict one environment for the hydrogens of the four terminal methyl groups and a different environment for the hydrogens on the two bridging methyl groups, with a ratio in the peak areas of $2:1$. When the temperature was lowered to $-75°C$, this pattern appeared. At intervening temperatures, the peaks coalesced into none-too-distinct shapes. These observations were interpreted as indicating "rapid exchange": the metal–carbon bonds break and reform so rapidly at room temperature that the NMR detector "sees" the two different environments as one single average environment. As the temperature is lowered, the rate of exchange decreases to the point where each separate proton environment becomes detectable. Such rapid exchange occurs for all metals showing three-centered metal–carbon bonds, and for many other organometals as well. Use of the temperature variance of proton NMR signals has produced much valuable information on such exchange reactions.[5]

Rates of alkyl/aryl exchange depend on other factors in addition to temperature. The metals involved play a role: in general, the more electronegative the metal, the slower the rates of its exchange reactions. The size of the alkyl group is important, particularly in bridging positions, where steric factors become important. If trimethylaluminum dimer is mixed with dimeric trialkylaluminum compounds with larger groups, mixed bridges can be observed, but the majority of bridging positions will be occupied by the small methyl group, with the proportion increasing as the size disparity increases. As the alkyl groups become larger and more branched, the strength of the metal-alkyl-metal bridge decreases and may disappear altogether. Activation energies for exchange reactions also increase with increasing size:[5] for example, the activation energy for exchange in the CH_3Li/C_2H_5Li system is only 11 kcal/mol, but in the bulkier $(CH_3)_3CLi/(CH_3)_3SiCH_2Li$ system, this increases to 24 kcal/mol. The solvent used also makes a difference; the rate of exchange between t-butyllithium and trimethylsilyl-methyllithium increases by a factor of twenty when toluene replaces cyclopentane as solvent.[5]

4.4 ORGANOLITHIUM COMPOUNDS

Methyllithium is unique among the alkyllithium compounds by being insoluble in hydrocarbons. It is a white solid that decomposes without melting at 250°C. Melting points for straight-chain alkyllithium compounds decrease as the size of the alkyl group increases, with the result that n-butyllithium is a liquid at room temperature and readily soluble in hydrocarbon solvents. Organolithium compounds are air-sensitive, although less so than corresponding sodium or potassium compounds, and must be handled under inert atmosphere or similar protective conditions. They must also be protected from water or other protonic liquids, which can cleave the lithium–carbon bond:

$$RLi + H_2NR' \longrightarrow LiNHR' + RH \qquad (4.1)$$

Ethers and tertiary amines can form complexes with organolithium compounds, destroying or lowering their association. For example, the following equilibrium[6]

$$(n\text{-}C_4H_9Li)_4 \cdot 4\ C_4H_8O + 4\ C_4H_8O \rightleftharpoons 2\ (n\text{-}C_4H_9Li)_2 \cdot 2\ C_4H_8O \quad (4.2)$$

has an activation energy of 41 kJ/mol and an activation entropy of -30 J/K·mol. In ethers there is always the possibility of C—O bond cleavage:

$$RLi + R'OR'' \longrightarrow LiOR'' + RR' \quad (4.3)$$

Even sufficiently acidic hydrocarbons can cause cleavage in a metalation reaction:

$$RLi + R'H \longrightarrow R'Li + RH \quad (4.4)$$

This reaction can be useful for preparing organolithium compounds containing exotic organic groups that might be difficult or impossible to prepare in other ways. The reagent of choice here is n-butyllithium, which is easily prepared, reasonably simply to handle in hydrocarbon solvents, and is the lithium salt of gaseous butane, one of the least acidic of the hydrocarbons.

Organolithium reagents are extensively used in organic synthesis, primarily to attach the organic group to some metalloid or to attach another functional group onto the organic moiety. Discussion of these applications are beyond the scope of this book. Organolithium compounds can also be used as catalysts for the polymerization of dienes and olefins, undergoing organometalation reactions similar to the better known organoaluminum compounds.[7]

The organolithium compounds heretofor mentioned have had one lithium atom per organic group. In principle, any and every hydrogen atom of a hydrocarbon might be replaced by lithium, and there have been some reports of polylithio compounds. Treatment of propyne, C_4H_4, with excess n-butyllithium yielded the tetralithio derivative, C_4Li_4, which, upon treatment with $(CH_3)_3SiCl$, gave the major product, $C_3[Si(CH_3)_3]_4$, in 75% yield:[7]

$$C_3H_4 + 4\ n\text{-}C_4H_9Li \longrightarrow C_3Li_4 + 4\ n\text{-}C_4H_{10} \quad (4.5)$$

$$C_3Li_4 + 4\ (CH_3)_3SiCl \longrightarrow C_3Si(CH_3)_3{}_4 + 4\ LiCl \quad (4.6)$$

The trisilylated derivative formed in 15% yield, and various minor products were detected by vapor phase chromatography. Interestingly enough, the major product was not the expected tetrasilylpropyne, but rather the isomeric allene $[(CH_3)_3Si]_2C{=}C{=}C[Si(CH_3)_3]_2$.

The potential synthetic utility of a polylithiohydrocarbon is shown in Figure 4.2. The compound 1-lithio-1-phenyl-2-(o-lithiophenyl)-1-hexene forms by the reaction of one mole of diphenylacetylene with two moles of n-butyl-lithium in ether.[8] This molecule can then react with various metal

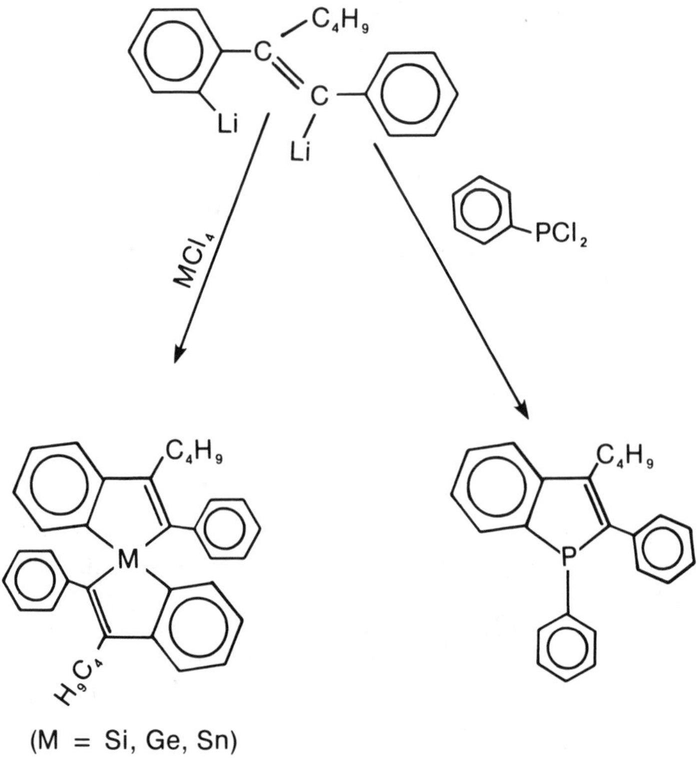

(M = Si, Ge, Sn)

Figure 4.2. Use of 1-lithio-2-phenyl-2-(o-lithiophenyl)-1-hexene for organometal synthesis.

or organo–metal halides to form metallocyclic products. The positioning of the two lithium atoms in the molecule is particularly favorable for the formation of cyclic products.

Recently there has been much interest in "hypervalent" organolithium compounds, which have a greater Li/C ratio than would be expected from standard bonding theory. Theoretical calculations indicate that species such as CLi_5 and CLi_6 would have a considerable degree of stability.[9] It should be remembered in this context that lithium is the smallest of the metals, and that Li^+ is both strongly electrophilic and smaller than carbon. This area will doubtless develop quite considerably in years to come.

4.5 ORGANOMAGNESIUM COMPOUNDS

Diorganomagnesium compounds in the solid state have a polymeric chain structure with bridging organic groups. Upon solvation, however, these bonds usually break, and molecules of type R_2ML_2 (L = donor solvent) form. The unsolvated compounds are reactive and relatively little studied.

Similarly, compounds of type RMgX are known primarily in solution (eg, the Grignard reagent), and will be discussed in the next chapter. Heating the dialkylmagnesium compounds in the absence of air causes elimination of olefins:

$$(C_2H_5)_2Mg \xrightarrow{\Delta} 2\ C_2H_4 + MgH_2 \qquad (4.7)$$

Also formed will be alkylmagnesium hydrides RMgH.

4.6 ORGANOBERYLLIUM COMPOUNDS

These compounds are less well known than their magnesium counterparts. Beryllium is smaller and more electronegative than either magnesium or lithium, with the result that steric effects become more significant (eg, di-*tert*-butylberyllium is monomeric, unlike the magnesium analog), and a greater number of anionic compounds (eg, Na_2BeR_4) are known. Investigations into organoberyllium chemistry have lagged due to the toxicity of the metal and the lack of any apparent laboratory or commercial applications. Nonetheless, enough results have been published to warrant a review.[10] Organoberyllium hydrides resemble the corresponding boron and aluminum compounds. Methylberyllium hydride-trimethylamine is actually a dimer, $[CH_3BeH \cdot N(CH_3)_3]_2$, with Be—H—Be three-center bonds.[10]

4.7 ORGANOALUMINUM COMPOUNDS[11]

Neutral organoaluminum compounds may contain one, two or three Al—C linkages and may be represented by the formulas $RAlX_2$, R_2AlX, and R_3Al respectively. In addition, some anionic species of formula R_4Al^-, are also known. Only the triorganoaluminum compounds have "electron-deficient" bonding, and then primarily for those alkyl and aryl groups where steric hindrance is not a problem. As with the other metals in this chapter, aluminum can interact with donor solvents, frequently breaking the Al—C—Al bridges:

$$[(C_2H_5)_3Al]_2 + 2\ (CH_3)_3N \longrightarrow 2\ (C_2H_5)_3Al \cdot N(CH_3)_3 \qquad (4.8)$$

Bases containing protonic hydrogens, however, can cleave the aluminum–carbon bond:

$$[(C_2H_5)_3Al]_2 + 2\ (CH_3)_2NH \longrightarrow 2\ (C_2H_5)_3Al \cdot HN(CH_3)_2 \qquad (4.9)$$

$$2\ (C_2H_5)_3Al \cdot NH(CH_3)_2 \longrightarrow [(C_2H_5)_2AlN(CH_3)_2]_2 + 2\ C_2H_6 \qquad (4.10)$$

Occasionally the intermediate 1:1 complex may be isolated; in most cases, however, decomposition occurs readily.

Organoaluminum compounds of formulas $RAlX_2$ or R_2AlX, where X is an atom of Groups VA–VIIA bonded to aluminum, will exist as dimers or higher polymers. The dimers will have the same basic structure as $(CH_3)_6Al_2$, except that the atom X, which will have at least one pair of electrons available for donation, always occupies the bridging position. Similarly polymeric species have the same basic atomic skeleton as $(R_2Mg)_x$, again with the atom X always in a bridging position. These compounds have an extensive chemistry, and many are commercially important. Since the aluminum–carbon bond is always a sigma bond, these compounds will be considered in the next chapter.

REFERENCES

1. Rundle, R. E., *J. Phys. Chem.*, **1957**, *61*, 45.
2. Wakefield, B. J., "The Chemistry of Organolithium Compounds," Pergamon Press: Oxford, 1974.
3. Goldstein, M. J.; Wenzel, T. T., *Helv. Chim. Acta* **1984**, *67*, 2029.
4. Hoffmann, E. G., *Trans. Farad. Soc.*, **1962**, *58*, 642.
5. Oliver, J. P., *Adv. Organomet. Chem.*, **1970**, *8*, 167.
6. Garrity, J. F.; Ogle, C. A., *J. Am. Chem. Soc.*, **1985**, *107*, 1805.
7. West, R.; Jones, P. C., *J. Am. Chem. Soc.*, **1969**, *91*, 6156.
8. Anon., *Chem. Eng. News* (Nov. 8, 1971), 26.
9. Reed, A. E.; Weinhold, F., *J. Am. Chem. Soc.*, **1985**, *107*, 1919.
10. Coates, G. E.; Morgan, G. L., *Adv. Organomet. Chem.*, **1971**, *9*, 195.
11. Mole, T., "Organoaluminum Compounds," Elsevier: Amsterdam, 1972.

Chapter 5

Metal–Carbon Sigma Bonds. I: The Reactive Metals

5.1 INTRODUCTION

Sigma bonds, involving a metal(loid) atom bonded to a carbon atom by a shared pair of electrons, represent the most common type of metal–carbon linkage. Examples are known for virtually every metal or metalloid in the Periodic Table. Due to the widespread occurrence of these bonds, they need to be divided into categories for convenient discussion:

Reactive Metals — Metals of fairly low electronegativity. This category would include all organo derivatives of Zn, Cd, Ga, most transition metals, plus many derivatives of Be, Mg and Al.

Heavy Metals — Metals of the last two rows of the Periodic Table, with relatively high electronegativities. This category would include organo derivatives of Hg, In, Tl, Sn, Pb, Sb, Bi, Pt and Au.

Metalloids — This category would include the organo derivatives of B, Si, Ge, P, As, Se, Te, and higher-valent Br and I.

Many of these elements form other types of metal–carbon bonds, as for example in $(CH_3)_6Al_2$, mentioned in the preceding chapter. Such metals mentioned in this chapter would be discussed solely in terms of their metal–carbon sigma bonds.

5.2 CERTAIN GENERAL PROPERTIES

Compounds containing sigma bonds between reactive metals and organic groups will show some properties in common, although obviously there will be considerable variations among these compounds. These properties include:

High Bond Reactivity — The metal–carbon bond is quite polar, making these compounds quite labile. Many of them are used as reagents.

Ready Complex Formation — The metals in this category almost always have empty orbitals available to interact with donor molecules. Often the resulting complexes are less reactive than the starting organometal.

Lack of Alkyl Bridging — The "peralkyls" of these metals are volatile and monomeric. This differentiates them from the metals of the preceding chapters. Alkyl bridging may occur in unstable reaction intermediates.

Susceptibility to Protonolysis — Due to the strong polarity of the metal–carbon linkage, these compounds can react with protonic acids to give cleavage.

5.3 THE GRIGNARD REAGENT

The Grignard Reagent is probably the most extensively investigated organometallic system. This reagent has received intense scrutiny and has been the subject of no small debate. Even now, the nature of this reagent cannot be regarded as being completely understood. However, enough has been discovered to enable a reasonable discussion of its properties.

The Grignard Reagent is always prepared in solution, and is not isolated in pure form. In the great majority of cases, the solvent will be diethyl ether (b.p. 34.5°C), which is readily available commercially, easily removed from the reaction if necessarily, and not too difficult to handle. Other solvents would be used in special situations, such as the following:

a. *More basic solvent is needed* — In this case, diethyl ether might be replaced by tetrahydrofuran, triethylamine, etc.
b. *Higher reaction temperature is needed* — In this case, a higher-boiling ether, such as di-n-butyl ether (b.p. 142°C) might be used.
c. *Product separation is difficult* — Certain products (eg, trimethylarsine, b.p. 52°C) have boiling points inconveniently close to that of diethyl ether. Here again, a higher-boiling ether might be used to facilitate separation.

The oxygen atoms of the solvent molecules interact strongly with magnesium, producing a considerable energy of solvation, which is the major reason for the stability of the Grignard Reagent in solution. The fact that this reagent is prepared and used in solution, along with its reactivity, makes it much more satisfactory for most preparations than the air-sensitive organozinc compounds that it replaced. Solutions of Grignard Reagents are available commercially, and, if carefully protected, may be kept in storage for long periods of time.

In organic chemistry textbooks, the Grignard Reagent is usually written as "RMgX," implying that it exists as a monomeric species. This is perfectly satisfactory for writing chemical equations or for stoichiometric calculations, but does not always reflect the true nature of the reagent. Research indicates that the Grignard Reagent may have several different forms co-existing in a complex equilibrium, as shown in Figure 5.1. The coordination number of the magnesium atom is almost always four. In addition to various covalent forms, conductivity studies indicate the presence of ionization.[1] Molecular weight studies show a varying degree of association, depending on the following factors:

a. *Solvent* — As the basicity of the solvent increases, the degree of association decreases.
b. *Concentration* — The degree of association increases as the concentration increases.
c. *Halogen* — Association decreases in the order $F > Cl > Br > I$.
d. *Organic Group* — Little effect is observed. However, extremely large groups will favor dissociation for steric reasons.

5.4 METAL–CARBON BOND LABILITY AND SYNTHETIC UTILITY

The Grignard Reagent has been used for the preparation of many organometal(loid)s, as described in a monograph.[2] It provides such a potent combination of bond reactivity and ease of preparation/handling that it is the first choice of a synthetic chemist for use in organometal(loid) prepa-

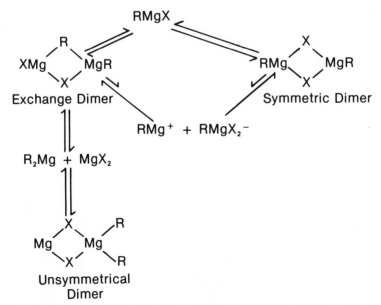

Figure 5.1. Various species found in Grignard Reagent solutions (solvating ether molecules not shown).

ration, and has become the standard against which other reagents are measured.

Certain systems may require a reagent that is more reactive than the Grignard. Usually, in such situations, an organolithium compound will be used. These are also commercially available and not too difficult to handle. They are a strong second to the Grignard Reagent in general use. For more extreme cases, organosodium or organopotassium compounds may be needed. These are virtually the most reactive organometal reagents available and are difficult to handle. They would be used for preparations where their great activity is necessary.

On the other hand, there are systems where the Grignard Reagent would be excessively reactive, and a milder reagent would be more appropriate. Grignard Reagents react with carbonyls and with carbon dioxide, giving addition across the carbon-oxygen double bond. Alkylzinc and alkylcadmium compounds do not so react; in fact, early reports on the use of alkylzinc compounds for synthesis frequently mentioned the use of carbon dioxide as a protective atmosphere. Alkylcadmium can be used to prepare ketones from acyl halides:

$$2 \underset{\underset{O}{\|}}{CH_3CCl} + (C_2H_5)_2Cd \longrightarrow 2 \underset{\underset{O}{\|}}{CH_3CC_2H_5} + CdCl_2 \qquad (5.1)$$

Under similar circumstances, a Grignard Reagent would react with the ketone to form the tertiary alcohol. Similarly, the gentler reactivity of alkylzinc compounds enables them to be used in the malonic ester synthesis or the Simmons–Smith reaction:

$$CH_2I_2 + Zn \longrightarrow ICH_2ZnI \qquad (5.2)$$

$$ICH_2ZnI + CH_3CH{=}CH_2 \longrightarrow \underset{H}{\overset{H}{CH_3C}} \underset{\underset{C}{\diagup \diagdown}}{\diagdown \diagup} \underset{H}{\overset{H}{C}}{-}H + ZnI_2 \qquad (5.3)$$

The alkylzinc or -cadmium reagents used for such syntheses no longer have to be isolated, as in earlier days; instead, the zinc or cadmium halide is added in stoichiometric quantities to the corresponding Grignard Reagent, and the resulting solution used for the synthesis. Nor is this approach limited to zinc or cadmium organo compounds. Organoaluminum compounds have occasionally been used for synthetic purposes. In recent years, certain organocopper compounds of formulations $LiCuR_2$ and $MgCu_2R_4$ have been used for synthetic purposes.[3,4] Methylcobalamin can be used to synthesize certain methylmetal compounds in water (see Chapter 13).

Thus a wide range of alkylmetals (and arylmetals) now stands available for use in organometal(loid) synthesis, enabling considerable flexibility of choice. In addition, the development of laboratory equipment and techniques in recent decades has enabled the preparation, isolation, and characterization of compounds thought "impossible" not so long ago.

5.5 USES OF ALKYLMETAL COMPOUNDS IN POLYMER FORMATION

The high lability of many alkylmetal compounds has led to their use in various industrial applications. A very important example of this is the use of organometalation in the stereospecific synthesis of polymers:

$$m\text{—}CH_3 + CH_3CH\text{=}CH_2 \longrightarrow (CH_3)_2CHCH_2\text{—}m \qquad (5.4)$$

The compound most commonly used for this purpose is triethylaluminum, although others can also be used. These processes were developed independently by Drs. Ziegler and Natta,[5-7] who shared the 1962 Nobel Prize in Chemistry for their work.

The reaction shown in equation 5.4 requires a catalyst. Many have been used, and others are constantly being developed. One such catalyst is titanium trichloride. In the solid state, this compound has a coordination number of six, and the chlorides bridge from one titanium atom to another. At the solid surface, however, many titanium atoms will have an empty coordination site, enabling attack by triethylaluminum in the following manner:

$$\bigcirc\text{—}Ti(\text{—}Cl\text{—})_5 + (C_2H_5)_3Al \longrightarrow C_2H_5\overset{\bigtriangleup}{Ti}(\text{—}Cl\text{—})_4$$
$$+ (C_2H_5)_2AlCl \qquad (5.5)$$

$$\bigcirc\text{—}Ti(\text{—}Cl\text{—})_4C_2H_5 + CH_2\text{=}CH_2 \longrightarrow$$
$$\underset{\overset{\|}{H_2C}}{H_2C}\text{—}Ti(\text{—}Cl\text{—})_4C_2H_5 \qquad (5.6)$$

$$\underset{\overset{\|}{H_2C}}{H_2C}\text{—}Ti(\text{—}Cl\text{—})_4C_2H_5 \longrightarrow \bigcirc\text{—}Ti(\text{—}Cl\text{—})_4C_4H_9 \qquad (5.7)$$

where $\bigcirc\text{—}Ti$ represents a vacant coordination site and $(\text{—}Cl\text{—})$ represents a bridging chloride group. The reaction shown in equation 5.7 ("ethyltitanation") represents the important portion of the cycle. Equations

5.6 and 5.7 can repeat themselves indefinitely, producing an alkyl chain that grows by two carbon atoms per cycle. For unsymmetric olefins, this addition occurs stereospecifically to produce the highly crystalline isotactic polymer.

This mechanism also illustrates a major difference between the transition metal titanium and the representative metal aluminum. In addition to being somewhat less electronegative, titanium can form a synergistic metal–carbon linkage (see Chapter 8 for a discussion of this type of bond) to olefin molecules, simultaneously binding them in close proximity to the labile Ti—C sigma bond and weakening the carbon–carbon pi-bond, making reaction easier. Thus, while triethylaluminum will undergo organometalation without a catalyst, the presence of titanium trichloride facilitates the reaction and enables more complete control of product formation.

5.6 SIGMA BONDED ORGANOTRANSITION METAL COMPOUNDS

Prior to the discovery of ferrocene, chemists generally believed that transition metals (except for platinum and gold, discussed in the next chapter) did not form stable sigma-bonded organometals. Since 1951, however, many stable alkyl or aryl derivatives of transition metals have been reported, and the reasons for the instability of others have become understood. Instability appears primarily due to kinetic factors, such as the following:

a. *Coordination Unsaturation* — The metal has less than the ideal number of ligands. Empty orbitals are thus available, facilitating reaction, as with $TiCl_3$ in the preceding section.
b. *Ready Dehydrometalation* — If the alkyl group has a hydrogen in the *beta* position, it can undergo ready elimination as an olefin.
c. *Sigma-pi Rearrangement* — Aryl or unsaturated groups bonded to a transition metal often undergo ready rearrangement to form synergistic metal–carbon linkages.

Empty orbitals in the coordination sphere of transition metals appear to be the primary factor in the ready decomposition of many sigma-bonded organometals. Titanium, for example, has a coordination number of six. Tetramethyltitanium decomposes at $-80°C$, but complexes of type $(CH_3)_4Ti \cdot Li_2$ (L = donor ligand) are isolable. Similarly, $(t\text{-}C_4H_9)_4Zr$ is very unstable, even at $-78°C$, but a bis-pyridine adduct is considerably more stable.[8] While CH_3Cu is ill-defined and quite unstable, Li and Mg salts containing CuR_2^- are stable enough to be used as reagents.[3,4] The same is true for organozirconium compounds, which are becoming increasingly important in organic synthesis.[9] Compounds containing both sigma and synergistic metal–carbon bonds, such as $CH_3Mn(CO)_5$, are also known. These are frequently important as intermediates in catalyzing organic rearrangements, and will be discussed in a later chapter.

Alkyl groups containing a β-hydrogen when bonded to a transition metal will undergo ready dehydrometalation:

$$m—CH_2CH_2CH_3 \longrightarrow m—H + CH_2{=}CHCH_3 \qquad (5.8)$$

As a result, many stable methylmetal compounds are known, as well as species such as $(t\text{-}C_4H_9)_4Cr$ or $(C_6H_5)_2CHCrCl_2{\cdot}2\ C_4H_8O$, but stable ethylmetal derivatives are virtually nonexistent. This elimination reaction is another reason that many transition metals are used as catalysts in organic rearrangements.

Aryl groups bonded to transition metals are usually unstable simply because they can rearrange to the more stable synergistic compounds.[10] The reaction between cyclopentadienyllithium and $FeCl_2$ at low temperatures gives a sigma-bonded derivative, but upon warming, this compound rearranges to ferrocene. Similarly, treatment of $CrCl_3$ with phenylmagnesium bromide gave $(C_6H_6)_2Cr$.

Transition metals generally have rather low electronegativities, usually comparable to magnesium or aluminum.[11] Thus the sigma-bonded alkyl groups will have a polar M—C linkage, which makes these compounds very reactive chemically. Placement of electron-withdrawing groups on the carbon atoms in place of hydrogen usually makes these bonds more stable. This has been most studied for fluorine, and many fluorocarbon derivatives of metals are known.[12] Frequently, there is a substantial change in stability: $CH_3Co(CO)_4$ decomposes at $-30°C$, but $CF_3Co(CO)_4$ boils unchanged at $91°C$. Similarly, the synergistic compound $(C_6F_6)_2Cr$ is unknown, while $(C_6F_5)_2Cr$ has been prepared—just the opposite of the hydrocarbon system!

In recent years some stable transition metal "peralkyls" have been reported. Hexamethyltungsten, $(CH_3)_6W$, prepared from WCl_6 and CH_3Li, is a red solid that melts at approximately $30°C$. It reacts with strong acids in a fashion similar to other active methylmetals:[13]

$$(CH_3)_6W + 6\ HCl \longrightarrow WCl_6 + 6\ CH_4 \qquad (5.9)$$

Pentamethyltantalum, $(CH_3)_5Ta$, is a yellow oil, decomposing at $25°C$, which forms a stable adduct[14] with $(CH_3)_2PCH_2CH_2P(CH_3)_2$. The corresponding niobium compound $(CH_3)_5Nb$ also formed an adduct with this ligand. Both compounds reacted with excess methyllithium to form the anionic species $(CH_3)_7M^{2-}$ in solution. Thermodynamic calculations indicated mean bond dissociation energies of 261 ± 6 kJ/mol for Ta—CH_3 and 159 ± 7 kJ/mol for W—CH_3 respectively.[15]

Compounds containing sigma-bonded groups and synergistically bonded groups on the same metal are becoming increasingly common. Many are known for zirconium,[9] and are being reported for some actinides as well[16] in the form of compounds such as $(Me_5C_5)_2Th(CH_3)_2$ and $(Me_5C_5)_3UC_6H_5$, where Me_5C_5 represents the pentamethylcyclopentadienyl group. The extensive and growing importance of these compounds as catalysts and synthetic reagents indicates that there is likely to be much more work done on them.

5.7 ORGANO COMPOUNDS OF ZINC, CADMIUM, AND GALLIUM

Zinc and cadmium form two series of organometallic compounds, having formulas R_2M and RMX respectively; the first series is much the better known. The simple dialkyls of zinc and cadmium are volatile colorless liquids, inflammable in air and reactive toward water or protonic solvents. They show no tendency to form isolable dimers or higher polymers containing bridging alkyl groups. However, these compounds do undergo rapid exchange with corresponding alkyls of magnesium (and with each other), and the exchange occurs through bridging alkyl intermediates.[17] Although alkylzinc compounds have been replaced for the most part by the Grignard Reagent for synthetic purposes, they still serve as useful reagents for the previously mentioned Simmons–Smith reaction and for other preparations.[18]

In appropriate solvents, the monoalkylzinc and monocadmium species RMX behave much like the Grignard Reagent, except that there is considerably less tendency to associate, and the monomer tends to be the predominant form. An equilibrium study on the $(CH_3)_2Cd/CdI_2$ system[19]

$$(CH_3)_2Cd + CdI_2 \longrightarrow 2\ CH_3CdI \qquad (5.10)$$

indicated that the equilibrium constant exceeded 110.

Structural investigations on RZnX (X = halide or an O- or N-containing ligand bonded to Zn) indicate tetrameric structures for such compounds in the solid state [eg, $(CH_3ZnOCH_3)_4$], having the Zn atoms and the bridging atoms at alternating corners of an irregular cube.

Organogallium compounds generally resemble the corresponding aluminum compounds in formal stoichiometry, with the major exception that alkyl bridging groups are much less common. Trimethylgallium exists as a monomer in solution,[17] although trivinylgallium is dimeric. Trialkylgallium compounds do undergo ready alkyl exchange, presumably through bridging alkyl intermediates. The trialkyls are basic, accepting carbanions to form anions [eg, $(CH_3)_4Ga^-$] and forming complexes with donor ligands; for example, $(CH_3)_3Ga \cdot O(C_2H_5)_2$ is an isolable adduct that boils at 98.3°C.

In general, organogallium compounds are less reactive than their aluminum counterparts, undergoing only partial hydrolysis:

$$(CH_3)_6Al_2 + 3\ H_2O \longrightarrow 6\ CH_4 + Al_2O_3 \qquad (5.11)$$

$$(CH_3)_3Ga + H_2O \longrightarrow 2\ CH_4 + 1/n[CH_3GaO]_n \qquad (5.12)$$

The lower reactivity of gallium, along with the scarcity and high cost of the metal and the lack of any particular synthetic usefulness, have hampered the development of organogallium chemistry. Much less is known about it than about the corresponding chemistry of aluminum, zinc, or magnesium.

REFERENCES

1. Ashby, E. C., *Quart. Rev.,* **1967,** *21,* 259.
2. Kharasch, M. S.; Reinmuth, O., "Grignard Reactions of Nonmetallic Substances," Constable & Company: London, 1954.
3. House, H. O., *Accts. Chem. Res.,* **1976,** *9,* 60.
4. Goel, A. B.; Ashby, E. C., *Inorg. Chim. Acta,* **1981,** *54,* L199.
5. Ziegler, K., *Angew. Chem.,* **1964,** *76,* 545.
6. Natta, G., *Angew. Chem.,* **1964,** *76,* 555.
7. Sinn, H.; Kaminsky, W., *Adv. Organomet. Chem.,* **1980,** *18,* 99.
8. Bougeard, P.; McCullough, J. J.; Sayer, B. G.; McGlinchey, M. J., *Inorg. Chim. Acta.,* **1984,** *89,* 133.
9. Negishi, E.; Takahashi, T., *Aldrichimica Acta,* **1985,** *18,* 31.
10. Tsutsui, M.; Courtney, A., *Adv. Organomet. Chem.,* **1977,** *16,* 241.
11. Huheey, J. E., "Inorganic Chemistry," Harper & Row: New York, 3rd. ed., 1983, pp. 144–160.
12. Treichel, P. M.; Stone, F. G. A., *Adv. Organomet. Chem.,* **1964,** *1,* 143.
13. Shortland, A.; Wilkinson, G., *J. Chem. Soc., Dalton Trans.,* **1973,** 872.
14. Schrock, R. R.; Meakin, P., *J. Am. Chem. Soc.,* **1974,** *96,* 5288.
15. Adedeji, F. A.; Connor, J. A.; Skinner, H. A.; Galyer, L.; Wilkinson, G., *J. Chem. Soc., Chem. Comm.,* **1976,** 159.
16. Marks, T. J., *Science,* **1982,** *217,* 989.
17. Oliver, J. P., *Adv. Organomet. Chem.,* **1970,** *8,* 167.
18. Furukawa, J.; Kawabata, N., *Adv. Organomet. Chem.,* **1974,** *12,* 83.
19. Cavanagh, K.; Evans, D. F., *J. Chem. Soc. (A),* **1969,** 2890.

Chapter 6

Metal–Carbon Sigma Bonds. II: The Heavy Metals

6.1 INTRODUCTION

The "heavy metals" considered in this chapter are those elements in the last two rows of the Periodic Table: indium, tin, antimony, mercury, thallium, lead, bismuth, platinum and gold. Certain compounds of silver and palladium would also fall into this category of organometals. As might be expected, the organo derivatives of these metals show frequent resemblances to the corresponding compounds of their lighter congeners. However, they also show sufficient similarities to each other to warrant being placed in a separate category. Such similarities include:

a. *Solubility in, and stability towards, protic solvents* — Previous classes of organometals react rapidly with acids to give metal–carbon bond cleavage. Compounds in this category show stability towards water and dilute acids, frequently with marked solubility.
b. *Homolytic metal–carbon bond cleavage* — Thermal or photolytic decomposition for these compounds occurs through homolytic cleavage, with formation of radicals. Bond energies are generally low, and dissociation can proceed under relatively mild conditions.
c. *Expansion of coordination number* — While known for other categories of organometals, expansion of the coordination shell is especially important for the heavy metals. Many complexes with mono- and bidentate ligands have been reported.
d. *Formation of cationic species* — Most metals in this category can form "onium" compounds of type R_nM^+, frequently by dissociation in polar solvents such as water.
e. *Different oxidation states* — Except for mercury, every metal in this category can form organo derivatives with the metal in either of two oxidation states. Much of the chemistry of these compounds involves interconversions between these states.

6.2 ORGANOMETALS CONTAINING METAL–METAL BONDS

Most compounds of this type have been reported for tin, lead, antimony and bismuth. The inorganic salts of Hg_2^{2+} and Hg_3^{2+} have not had any organo counterparts yet reported, nor are there any reported organoindium or organothallium compounds. A few ylid derivatives containing Au—Au bonds have been reported.[1] There seems no intrinsic reasons for these bonds to be unstable; doubtlessly, as synthetic and characterization techniques become steadily more sophisticated, such compounds will be reported. Organometals containing two different metals bonded to each other are also known and are becoming increasingly common.

Polytins (polystannanes) are the best known and characterized compounds of this type. There are two series: Sn_xR_{2x+2}, corresponding to the alkane series of hydrocarbons, and $(SnR_2)_x$, corresponding to the cycloalkanes; specific examples include $(CH_3)_8Sn_3$, octamethyltritin (or octamethyltristannane), and $(CH_3)_{10}Sn_5$, decamethylcyclopentastannane. Recently, the cyclostannane, R_6Sn_3 (R = 9-phenanthryl-), was reported.[2] Organotin compounds with tin bonded to another metal are also known:

$$(C_2H_5)_2Zn + (C_6H_5)_3SnH \longrightarrow [(C_6H_5)_3Sn]_2Zn + 2\,C_2H_6 \quad (6.1)$$

Virtually all reported reactions of compounds containing Sn bonded to other metal atoms involve cleavage of the tin–metal bond.

The majority of compounds containing lead–lead bonds that have been characterized are the hexaorganodileads (hexaorganodiplumbanes), R_6Pb_2. Some octaorganotrilead compounds have been recently prepared:[3]

$$2\,R_3PbLi + R_2PbX_2 \longrightarrow R_8Pb_3 + 2\,LiX \quad (6.2)$$

This reaction is complicated by the tendency of the triplumbane to react further with R_3PbLi; the product must be frozen out from solution in order to be isolated. The compound $(C_6H_5)_{12}Pb_5$, or $[(C_6H_5)_3Pb]_4Pb$, has also been reported.[4] Compounds containing organolead fragments bonded to other metals are more common, and include such species as cis-$(C_6H_5)_3Pb(C_6H_5)Pt[(P(C_6H_5)_3]_2$, containing both Pb—C and Pt—C bonds.[5] In a typical preparation of such compounds, $(C_6H_5)_3PbLi$ reacted readily with triphenyltin chloride.[6]

$$(C_6H_5)_3PbLi + (C_6H_5)_3SnCl \longrightarrow (C_6H_5)_3PbSn(C_6H_5)_3 \quad (6.3)$$

and with the corresponding germanium compound, but not with triphenylchlorosilane. Treatment of triphenylsilyllithium with triphenyllead chloride gave an equimolar mixture of hexaphenyldisilane and hexaphenyldilead.[6]

The best known compounds containing antimony–antimony linkages are the tetraorganodistibines, R_4Sb_2. Cyclic polystibines of type $(RSb)_x$ are also known. Hexaphenylcyclohexastibine, $(C_6H_5Sb)_6$, prepared by careful oxidation of $C_6H_5Sb[Si(CH_3)_3]_2$, was found to have the chair conformation, with all phenyl rings in equatorial positions.[7]

Tetraphenyldibismuth, $(C_6H_5)_4Bi_2$, an air-sensitive material, has been prepared by reaction of iododiphenylbismuth and sodium in liquid ammonia,[8] and its crystal structure determined.[9] The methyl analog reacts readily with diphenylsulfide (also the selenium and tellurium analogs):[10]

$$(CH_3)_4Bi_2 + 2\ (C_6H_5)_2S \longrightarrow [2\ (CH_3)_2BiSC_6H_5] \longrightarrow$$
$$CH_3Bi(SC_6H_5)_2 + (CH_3)_3Bi \quad (6.4)$$

An unusual compound containing a Bi—Bi bond forms from the dimerization of 4-*tert*-butyl-1-bismabenzene.[11] Bismuth can form bonds to other metals as well.

Almost every metal–metal bond discussed has been a single bond. The compound $[\{(CH_3)_3Si\}_2CH]_2Sn=Sn[CH\{Si(CH_3)_3\}_2]_2$ has been reported, with a bond length somewhat shorter than a tin–tin single bond length.[12] Double bonds between these metals and some lighter elements are also known,[12] most notably antimonin and bismin, $C_5H_5M(M = Sb, Bi)$, analogous to pyridine. Like the phospha and arsa counterparts, these compounds have some aromatic character (see Chapter 7).

The instability of metal–metal bonds for these elements appears to be kinetic rather than thermodynamic, with the availability of low-energy pathways to decomposition or rearrangement facilitating decomposition. As techniques improve, most likely additional compounds of this type will be isolated and characterized, and presently unknown compounds (eg, $CH_3Hg—HgCH_3$, $(C_6H_5)_2Tl—Tl(C_6H_5)_2$, etc.) may also be reported.

6.3 ORGANO DERIVATIVES OF METALS IN LOWER OXIDATION STATES

Every metal discussed in this chapter has inorganic compounds in at least two different oxidation states. Some of these, such as mercurous nitrate, $Hg_2(NO_3)_2$, or distannane, Sn_2H_6, have metal–metal bonds. Others have mononuclear compounds in lower oxidation states; these are actually the more abundant and stable compounds for thallium, lead and bismuth. By contrast organo derivatives are most abundant for these metals in their higher oxidation states; only antimony, bismuth, platinum and gold have an extensive organic chemistry in two oxidation states.

Mercury forms only divalent organo derivatives. Indium and thallium form cyclopentadienyl (and substituted cyclopentadienyl) derivatives of formula C_5H_5M. These compounds are polymeric solids whose structures show alternating metal atoms and cyclopentadiene rings in endless stacks. Tin and lead are predominantly quadrivalent in their organo chemistry. Attempts to prepare species of type R_2M by reaction of MX_2 with RMgX gives cyclic polymeric $(R_2Sn)_x$ for tin or R_4Pb plus Pb for lead. Some stable organo derivatives of Sn(II) and Pb(II) have been isolated.[13] "Stannocene" and "plumbocene," $(C_5H_5)_2M$, are monomeric compounds, in contrast to the corresponding isoelectronic In(I) and Tl(I) compounds. They can form 1:1 complexes with BF_3 to form polymers having bridging fluorine atoms. Carborane derivatives containing M(II)–carbon bonds are known,[13] as is the compound $[\{(CH_3)_3Si\}_2CH]_2Sn$. This compound can form complexes with transition metals of formula $R_2Sn—Mo(CO)_5$, indicating a chemical similarity to the isoelectronic species R_3Sb. As more organotin(II) and organolead(II) compounds are isolated and characterized, they should add considerably to the organo chemistry of these metals.

The organo chemistry of antimony(III) and bismuth(III) will be discussed in the section on those two metals; the same is true for gold(I) and platinum(II). Much of the chemistry of these series of compounds relates to their abilities to undergo oxidation.

6.4 ORGANOMERCURY COMPOUNDS

Organomercury chemistry is extensively developed and has a long history. These compounds were used in the late 19th Century in medicine, and continue to be used to a certain extent. They have considerable importance in organic syntheses, and have received much renewed attention in recent years due to their toxicity and implications in environmental poisonings. The biological effects and uses of organomercurials will be discussed in Chapters 11–13.

Neutral organomercury compounds may have two Hg—C linkages, as in R_2Hg, or one, as in RHgX. The former are covalent compounds, often with considerable volatility. Dimethylmercury, $(CH_3)_2Hg$, is a colorless liquid (B.P. 93–6°C) that is insoluble in water but readily soluble in hydrocarbons. By contrast, monoorganomercurials are usually solids. If the inorganic group is a weak base (eg, NO_3^-), the organomercury compound may be ionic; most, however, are covalent. Methylmercuric chloride, for example, is a covalent solid, having some volatility at room temperature and readily soluble in organic solvents. The hydroxides, RHgOH, react with strong acids:

$$C_2H_5HgOH_{(s)} + HCl_{(aq)} \longrightarrow C_2H_5HgCl_{(s)} + H_2O_{(l)} \qquad (6.5)$$

Methylmercuric compounds show some solubility in water and have an extensive aqueous chemistry.[14] The ion CH_3Hg^+ reacts with a variety of nitrogen, oxygen or sulfur donor species, and is often used as a reference acid in comparative acid–base investigations.[14]

The mercury–carbon bond is rather weak thermodynamically, and unreactive towards oxygen. It is labile enough, however, to allow ready exchange:

$$RHg^+ + Hg^*X^+ \longrightarrow R^*Hg^+ + HgX^+ \qquad (6.6)$$

An important rearrangement involving mercury–carbon bonds has recently been reported:

$$2\,CH_3Hg^+ + 2\,S^= \longrightarrow (CH_3Hg)_2S \qquad (6.7)$$

$$(CH_3Hg)_2S \xrightarrow{\Delta} (CH_3)_2Hg + HgS \qquad (6.8)$$

This reaction, which is also known for selenium, is important in the biological detoxification of methylmercuric compounds, and is discussed further in Chapter 12.

The low activity of mercury makes diorganomercurials useful for preparing other organometals (see Chapter 2). Mercuric acetate reacts with benzene and other aromatic hydrocarbons to form mercury–carbon linkages, and undergoes an oxymercuration reaction with olefins:

$$ClHgOH + H_2C{=}CH_2 \longrightarrow HOCH_2CH_2HgCl \qquad (6.9)$$

An unusual organomercurial is tetrakis(acetoxymercuri)methane,

$$\overset{O}{\overset{\|}{(CH_3COHg)_4C}}$$, in which four mercury atoms are bonded to one carbon. This molecule, which contains four electron-rich mercury atoms in a rather small volume of space, binds readily to coat proteins of bacteriophages and makes an excellent imaging agent for electron microscopy.[15]

6.5 ORGANOINDIUM AND ORGANOTHALLIUM COMPOUNDS

The chemistry of these two metals is probably the least developed of the heavy metals; what is known suggests some similarity to the lighter congenors aluminum and gallium, but more similarity to other heavy metals.[16] Both metals form compounds containing one, two or three metal–carbon linkages per metal atom, and a few anionic species of type R_4M^- are known. Organothallium chemistry has been reviewed.[17]

Treatment of the metal(I) halides with Grignard reagent gives the trialkylmetal:

$$3\,MBr + 3\,RMgBr \longrightarrow R_3M + 2\,M + 3\,MgBr_2 \qquad (6.10)$$

The corresponding reaction with $TlCl_3$ produces only R_2TlCl.

Both metals form many species containing the R_2M^+ ion; in fact, these derivatives are the most stable and common organothallium compounds.

Depending on the organic group, and to some extent on the anion, the diorganothallium compounds can show considerable salt-like character and dissolve appreciably in water. The hydroxides, R_2TlOH, react readily with acids, providing a convenient route for the preparation of various R_2TlY derivatives. Dimethylthallium ion has a linear C—Tl—C framework and is isoelectronic with dimethylmercury. Trimethylthallium reacts with water to form this ion and methane. Dialkylindium ions are known, but do not have the dominance of the corresponding thallium species.

Monoalkylmetal compounds are relatively rare. The series "$RInI_2$" (R = methyl, ethyl, n-butyl) actually has the structure $R_2In^+ InI_4^-$, and exists as dimers in solution.[18] Methylthallium diacetate may be prepared from the corresponding dimethylthallium compound:

$$(CH_3)_2TlC_2H_3O_2 + Hg(C_2H_3O_2)_2 \longrightarrow$$
$$CH_3Tl(C_2H_3O_2)_2 + CH_3HgC_2H_3O_2 \quad (6.11)$$

This compound undergoes ready decomposition in the presence of aqueous halide:

$$CH_3Tl(C_2H_3O_2)_2 + 2\ Cl^- \longrightarrow TlCl + CH_3Cl + 2\ C_2H_3O_2^- \quad (6.12)$$

The instability of organothallium dihalides towards reductive elimination presumably is the reason for this reaction. Monoorganometal compounds are even more unstable for lead and bismuth. They are better characterized for indium.[19]

Compounds containing metal–metal bonds are quite rare for these elements, although this may be due to lack of any concerted effort to prepare them. Triphenylindium can act as a Lewis base, and forms complexes[19] such as $(C_6H_5)_3InMn(CO)_5^-$. The only reported species with a Tl—Tl bond is $(CH_3)_6Tl_2^{2-}$, and even this has been questioned.[20]

Thallium(III) acetate undergoes a variety of oxythallation reactions. In most cases, the organothallium compound is unstable and decomposes; for this reason, thallium(III) acetate has become a very useful reagent for certain organic syntheses. In some cases, the organothallium product can be isolated:[20]

$$C_6H_5C\equiv C_2H_5 + Tl(C_2H_3O_2)_3 \longrightarrow$$
$$trans\text{-}C_2H_3O_2(C_6H_5)C{=}C(C_2H_5)Tl(C_2H_3O_2)_2 \quad (6.13)$$

6.6 ORGANOTIN COMPOUNDS

Organotin compounds have become the most important commercial organometal among the heavy metal species. They are widely used as plastic stabilizers, wood preservatives, water repellants, antitumor agents, insect repellants, etc. One estimate states that 30,000 to 35,000 tons of organotin compounds are used each year.[21]

Mononuclear organotin compounds may have one, two, three or four tin–carbon bonds. While many examples of each type have been reported, the triorganotin compounds, R_3SnX, are the most common. The Sn—X linkage will have a certain degree of ionic character, depending on both X and R; for example, the melting points of the trimethyltin halides vary in the order: fluoride (>300°C) > chloride (37°C) > bromide (27°C) > iodide (3.4°C). X-ray crystal studies on $(CH_3)_3SnF$ indicated a trigonal bi-pyramidal structure (see Figure 6.1) with the Sn and the three methyl carbon atoms in the same plane, and bridging fluorides.[21] Trimethyltin chloride, by contrast, is a covalent solid, having a marked vapor pressure (and unpleasant odor!) at ambient temperatures. Other triorganotin compounds are solids, with reasonable solubilities in hydrocarbons and some solubility in water. The tendency towards hydrolysis is much less than for the corresponding Si or Ge compounds, and tin–oxygen bonds can react with aqueous acids.

Diorganotin compounds usually show greater bridging and polymeri-zation in the solid state, especially if oxygen or nitrogen is present. Hy-drolysis of diorganotin dihalides occurs in stages:

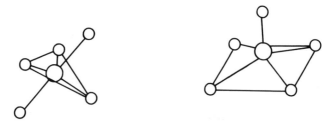

Trigonal bipyramidal structure   Square pyramidal structure

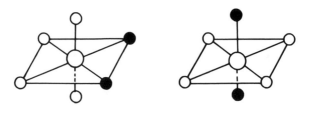

Octahedral Structures

Cis   Trans

Figure 6.1. Geometrical structures of coordination numbers 5 and 6.

$$2 \; R_2SnCl_2 + H_2O \longrightarrow \underset{\underset{Cl}{|}}{R_2Sn} - O - \underset{\underset{Cl}{|}}{SnR_2} + 2 \; HCl \tag{6.14}$$

The initial product can undergo further reaction, leading ultimately to polymeric R_2SnO. Monoorganotin compounds are the least well known organotin series and show even greater tendency to condense.

Numerous complexes between organotin compounds and donor molecules or ions have been reported.[22] These usually occur for the di- or triorganotin species, and involve expansion of the coordination number to five or six. In a compound such as $(CH_3)_3Sn(bipy)_2 \; {}^+Cl^-$ (bipy = 2,2'-bipyridine), the structure is a trigonal bipyramid, with the two nitrogen atoms of the bipyridine occupying adjacent equatorial coordination sites. Complexes of formula $(CH_3)_2SnX_2L_2$ have octahedral structures, with the C—Sn—C framework linear (ie, *trans*-octahedral), and both X groups and L groups in *cis* positions (see Figure 6.1). Unlike lead or mercury, tin apparently bonds more strongly to nitrogen or oxygen than to sulfur. Anionic complexes, such as $R_3SnX_2^-$, are also known. Such complexes play a role in the biological effects of organotin compounds.

Tetraorganotin compounds are completely covalent in nature, and show no solubility in water. They will react with strong acids to cleave the tin–carbon bonds:

$$R_4Sn + HCl \longrightarrow R_3SnCl + RH \tag{6.15}$$

More rigorous conditions are required for further cleavage. The tin–carbon bonds are labile, and redistribution occurs readily. The equilibrium

$$R_4Sn + R_4'Sn \rightleftharpoons R_3R'Sn + R_2R_2'Sn + RR_3'Sn \tag{6.16}$$

lies well to the side of the symmetric compounds; mixed tetraorganotins tend to be unstable towards this type of redistribution. As with mercury (see equations 6.7 and 6.8), trimethyltin ion undergoes redistribution in the presence of sulfide:[23]

$$2 \; (CH_3)_3SnCl + S^= \longrightarrow [(CH_3)_3Sn]_2S + 2 \; Cl^- \tag{6.17}$$

$$3 \; [(CH_3)_3Sn]_2S \longrightarrow [(CH_3)_3SnS]_3 + (CH_3)_4Sn \tag{6.18}$$

This reaction may well play an important role in the environmental circulation of methyltin compounds.

6.7 ORGANOLEAD COMPOUNDS

The commercial use of tetramethyllead and tetraethyllead as gasoline additives for much of the 20th Century has provided a considerable impetus for the development of organolead chemistry. In general it is quite similar to that of tin, with the two differences being the much readier reduction of Pb(IV) to Pb(II) and the greater lability of the Pb—C bond. Monoalkyllead compounds have never been isolated; monoaryllead carboxylates are stable, but react readily with aqueous halides:

$$ArPb(O_2CR)_3 + 3\ Br^- \longrightarrow ArBr + PbBr_2 + 3\ RCO_2^- \quad (6.19)$$

Di- and triorganolead halides also have a tendency towards reductive elimination, but this tendency decreases as the number of organic groups on lead increases.

Like the corresponding tin compounds, tetraorganoleads are covalent materials, soluble in hydrocarbons but not water, often volatile, sensitive towards acids, and can undergo exchange reactions. They have occasionally been used as alkylating agents in synthesis.

Di- and trialkyllead compounds generally resemble their tin counterparts in physical properties, except for their greater lability. The sulfide-induced rearrangement shown in equations 6.17 and 6.18 is also known for trimethyllead sulfide. Some alkyllead complexes have been reported, but considerably fewer than for the tin counterparts.

As with mercury(II) acetate and thallium(III) acetate, lead(IV) acetate reacts with olefins. The initially formed monoorganolead compounds are unstable and decompose readily:

$$(6.20)$$

$$(6.21)$$

For this reason, lead tetraacetate has become a very useful reagent for the introduction of oxygen-bound functional groups onto olefins. Lead tetraacetate also reacts with benzene and other aromatic hydrocarbons:

$$Pb(OAc)_4 + C_6H_6 \longrightarrow C_6H_5Pb(OAc)_3 + HOAc \quad (6.22)$$

6.8 ORGANOANTIMONY AND ORGANOBISMUTH COMPOUNDS

Two separate series of organoantimony and organobismuth compounds have been reported, having the metals in different oxidation states:

M(III): R_3M, R_2MX, RMX_2

M(V): $(C_6H_5)_6Sb^-$, Ar_5M, R_4M^+, R_3MX_2, R_2SbX_3, $RSbO_3H_2$

Both series have comparable importance for antimony, but the M(III) series is considerably more important for bismuth. The relative instability of Bi(V) makes compounds in this oxidation state quite labile, and many organo-antimony(V) derivatives have no bismuth counterparts. The early development of organoarsenic chemistry that arose from the medical work of Ehrlich (see Chapter 1) stimulated the corresponding chemistries of antimony and bismuth. As a result, both metals have an extensive organometallic chemistry.[24-26]

Triorganoantimony compounds, like their phosphorus and arsenic counterparts, have donor properties, and will form complexes with transition metal compounds. Since the antimonials are weaker bases, fewer such complexes are known than for the lighter Group VA elements. Triorganoantimony compounds can also act as Lewis bases under appropriate conditions:

$$2 \ CH_3SbCl_2 + SO_2Cl_2 \xrightarrow[-70°C]{acetylacetone} 2 \ CH_3SbCl_3(acac)^- + SO_2^{++} \quad (6.23)$$

The structure of this acetylacetonate complex is shown in Figure 6.2, and reflects the strong tendency of antimony to attain a coordination number of six in its compounds—a tendency that appears elsewhere in its chemistry. Triorganobismuth compounds show no donor properties, but can act as acceptors, and a few organobismuth(III) complexes are known.[26]

Triorganoantimony(III) compounds are readily oxidized:

$$R_3Sb + X_2 \longrightarrow R_3SbX_2 \quad (6.24)$$

$$R_3Sb + CH_3I \longrightarrow R_3SbCH_3^+ \ I^- \quad (6.25)$$

While triarylbismuth(III) compounds can be oxidized by halogens, trialkylbismuth(III) compounds undergo bismuth–carbon bond cleavage instead. As a result, almost all organobismuth(V) compounds have aryl groups attached to bismuth.

Pentaphenylbismuth and pentamethylantimony both have trigonal bipyramidal structures, as do the pentaphenyl derivatives of phosphorus and arsenic.[26] By contrast, pentaphenylantimony has a square pyramidal structure. Pentaphenylbismuth has an intense violet color, in contrast to the colorless antimony analog. Pentaphenylantimony can react with phenyllithium:

$$(C_6H_5)_5Sb + LiC_6H_5 \longrightarrow Li^+ \ (C_6H_5)_6Sb^- \quad (6.26)$$

Treatment of $(n-C_4H_9)_3SbBr_2$ with excess CH_3MgI in tetrahydrofuran at 0°C gave the expected $(n-C_4H_9)_3Sb(CH_3)_2$ in 75% yield, along with redistribution products $(n-C_4H_9)_nSb(CH_3)_{5-n}$ (n = 0 − 2) in lesser yields.[27]

Various tetraorganostibonium salts are known, including the optically active compound phenylmethylethylisopropylstibonium iodide. These salts, like their phosphonium and arsonium counterparts, can be used to precipitate large anions from solution. Tetraorganobismuthonium salts are

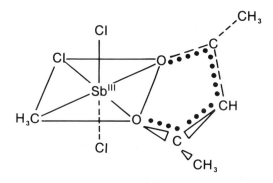

Structure of Methylantimonytrichloride acetylacetonate

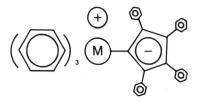

Structure of triphenylcyclopentadienylmetal cation(M = Sb, Bi)

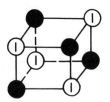

Structure of trimethylplatinum iodide tetramer(● = Me₃Pt)

Figure 6.2. Structures of some heavy metal species.

rare, and most of those reported contain the tetraphenylbismuthonium ion, such as $(C_6H_5)_4Bi^+$ $(C_6H_5)_4B^-$, melts without decomposition at 225°C. Treatment of triphenylbismuth dichloride with silver perchlorate in acetone gave the salt $[(C_6H_5)_3BiCH_2C(:O)CH_3]^+$ ClO_4^-, virtually the only example of an alkyl group bonded to Bi(V). The unusual compounds, $(C_6H_5)_3M—C_5(C_6H_5)_4$ shown in Figure 6.2, were prepared from the triphenylmetal compound and the diazonium salt of 2,3,4,5-tetraphenyl-cyclopentadiene.[28] These compounds may have some of the properties of ylids (see Chapter 7).

The metal (V) compounds containing three organic groups (aryls for bismuth) are the most numerous and best known for both antimony and bismuth. Halides attached to these metals are labile and exchange readily, although in the case of bismuth, reductive elimination may occur:

$$(C_6H_5)_3BiCl_2 + 2\ CN^- \longrightarrow (C_6H_5)_2BiCN + C_6H_5CN + 2\ Cl^- \quad (6.27)$$

Corresponding compounds with only one or two organic groups attached to the metal are unknown for bismuth and rare for antimony, due to their ability to undergo reductive elimination. Arylstibonic acids, $ArSbO_3H_2$, are the most stable examples of monoorganoantimony compounds. Recently reported are the antimonins and bismins (stiba- and bismabenzenes), C_5H_5M. These are analogs of pyridine, and like the better-known phospha- and arsabenzenes (discussed in Chapter 7) show some aromatic character in the ring system.

6.9 ORGANOPALLADIUM, ORGANOPLATINUM AND ORGANOGOLD COMPOUNDS

All three metals form synergistic metal–carbon bonds, and such bonds in fact dominate the organic chemistry of palladium. Nonetheless, many compounds containing metal–carbon sigma bonds are known for palladium,[29] platinum[30] and gold.[31] Each metal has two oxidation states that can exist in their organo derivatives: +2 and +4 for palladium and platinum; +1 and +3 for gold. The intermediate oxidation states (+3 and +2 respectively) are found in compounds with metal–metal bonds, although a mononuclear Pt(III) compound, $(n\text{-}C_4H_9)_4N^+ Pt(C_6Cl_5)_4{}^-$, has been reported.[32]

Organo compounds of Pd(II) and Pt(II) exist almost entirely as complexes, such as cis-$[(CH_3)_2PtP(C_2H_5)_3]_2$. Much of their chemistry consists of metal–carbon bond cleavage or rearrangements, and they are frequently found in molecules that also have metal–carbon synergistic bonds. Oxidation to M(IV) is known, mostly for platinum:

$$(CH_3)_2Pt[P(C_2H_5)_3]_2 + Cl_2 \longrightarrow (CH_3)_2PtCl_2[P(C_2H_5)_3]_2 \quad (6.28)$$

$$\text{trans-}CH_3PtI[P(C_2H_5)_3]_2 + CH_3I \longrightarrow (CH_3)_2PtI_2[P(C_2H_5)_3]_2 \quad (6.29)$$

Pt(II) and Pd(II) have square planar geometries, so that compounds of the type R_2ML_2 can be either *cis* or *trans*. Gold(I) also exists primarily as complexes, including the recently reported[33] $(C_6H_5)_3P\cdot AuCH_2C(:O)\text{-}C_5H_4Mn(CO)_3$; these have a linear arrangement of the atoms (eg, P—Au—C). Some species with two Au—C linkages are also known, such as $Li^+ Au(CH_3)_2{}^-$.

Very few Pd(IV) organo derivatives are known, reflecting the instability of this oxidation state relative to Pd(II). Those that have been reported include $(C_6F_5)_2PdCl_2\cdot bipy$ and $C_6F_5PdCl_3\cdot bipy$, and have an octahedral geometry. Organoplatinum(IV) compounds are much more numerous. The earliest and one of the most studied is $[(CH_3)_3PtI]_4$, whose structure is shown in Figure 6.2. All organoplatinum(IV) compounds have a coordination number of six around the platinum atom, and the resulting octahedral geometry may occasionally lead to unexpected structures.[30] Neutral tetramethylplatinum is unknown; reaction of trimethylplatinum iodide with methyllithium gave $Li_2Pt(CH_3)_6$ instead.[34] Many organoplatinum(IV) complexes are also known, such as Li^+ $(CH_3)_5PtP(C_6H_5)_3{}^-$ and $(CH_3)_3Pt(NH_3)_3{}^+$. While compounds derived from $[(CH_3)_3PtI]_4$ are the most common organoplatinum compounds, triethylplatinum iodide has been reported, and a number of dimethylplatinum(IV) compounds are also known.[30]

Organogold(III) compounds are less numerous than their platinum counterparts. They uniformly have a coordination number of four (it might be noted that Au(III) and Pt(II) are isoelectronic) and a square planar geometry. This occasionally causes polymerization; dimethylgold bromide is actually $[(CH_3)_2AuBr]_2$, with a four-centered rectangle of Au and Br. Trimethylgold is so unstable that it has never been isolated in pure form, but complexes such as $(CH_3)_3AuP(C_2H_5)_3$ or $Au(CH_3)_4{}^-$ Li^+ are stable at room temperature.

REFERENCES

1. Schmidbaur, H., *Accts. Chem. Res.*, **1975,** *8*, 62.
2. Fu, J.; Neumann, W. P., *J. Organometal. Chem.*, **1984,** *272*, C5.
3. Dräger, M.; Kleiner, N., *Abstracts of the Fourth International Conference on the Organometallic and Coordination Chemistry of Ge, Sn and Pb*, (1983), p. 7.
4. Willemsens, L. C.; van der Kerk, G. J. M., *J. Organometal. Chem.*, **1964,** *2*, 260.
5. Al-Allaf, T. A. K.; Butler, G.; Eaborn, C.; Pidcock, A., *J. Organometal. Chem.*, **1980,** *188*, 335.
6. Kleiner, N.; Dräger, M., *J. Organometal. Chem.*, **1984,** *270*, 151.
7. Breunig, H. J.; Haeberle, K.; Dräger, M.; Severengiz, T., *Angew. Chem.*, **1985,** *97*, 62.
8. Calderazzo, F.; Poli, R.; Pelizzi, G., *J. Chem. Soc., Dalton Trans.*, **1984,** 2365.
9. Calderazzo, F.; Morvillo, A.; Pelizzi, G.; Poli, R., *J. Chem. Soc., Chem. Comm.*, **1983,** 507.
10. Wiberg, M.; Sauer, I., *Z. Naturforsch.*, **1984,** *39B*, 1668.
11. Ashe, A. J.; Diephouse, T. R.; El-Sheikh, M. Y., *J. Am. Chem. Soc.*, **1982,** *104*, 5693.
12. Cowley, A. H., *Accts. Chem. Res.*, **1984,** *17*, 386.
13. Connolly, J. W.; Hoff, C., *Adv. Organomet. Chem.*, **1981,** *19*, 123.
14. Rabenstein, D. L., *Accts. Chem. Res.*, **1978,** *11*, 100.
15. Lipka, J. J.; Lippard, S. J.; Wall, J. S., *Science*, **1979,** *206*, 1419.
16. McKillop, A.; Smith, J. D.; Worrall, J. J., "Organometallic Compounds of Aluminum, Gallium, Indium and Thallium," Chapman & Hall: London, 1984.
17. McKillop, A.; Taylor, E. C., *Adv. Organomet. Chem.*, **1973,** *11*, 147.
18. Poland, J. S.; Tuck, D. G., *J. Organometal. Chem.*, **1972,** *42*, 315.
19. Tuck, D. G., in "Comprehensive Organometallic Chemistry," Pergamon: London, 1982, Vol. 1, pp. 684–723.
20. Kurosawa, H., in "Comprehensive Organometallic Chemistry," Pergamon: London, 1982, Vol. 1, pp. 684–723.
21. Davies, A. G.; Smith, P. J., in "Comprehensive Organometallic Chemistry," Pergamon: London, 1982, Vol. 2, pp. 519–627.
22. Petrosyna, V. S.; Yashina, N. S.; Reutov, O. A., *Adv. Organomet. Chem.*, **1976,** *14*, 63.
23. Craig, P. J.; Rapsomanikis, S., *J. Chem. Soc., Chem. Comm.*, **1982,** 114.
24. Okawara, R.; Matsumura, Y., *Adv. Organomet. Chem.*, **1976,** *14*, 187.
25. Freedman, L. D.; Doak, G. O., *Chemical Rev.*, **1980,** *82*, 15.
26. Wardell, J. L., in "Comprehensive Organometallic Chemistry," Pergamon: London, 1982, Vol. 2, pp. 681–707.

27. Nesmeyanov, A. N.; Borisov, A. E.; Novikova, N. V.; Fedin, E. I.; Petrovskii, P. V., *Chem. Abstr.*, **1974**, *80*, 48115.
28. Lloyd, D.; Singer, M. C., *Chem. Commun.*, **1967**, 1042.
29. Maitlis, P. M.; Russell, M. J. H., in "Comprehensive Organometallic Chemistry," Pergamon: Oxford, 1982, Vol. 6, pp. 279–350.
30. Hartley, F. R., *ibid.*, Vol. 6, pp. 471–762.
31. Puddephat, R. J., *ibid.*, Vol. 2, pp. 765–821.
32. Uson, R.; Fornies, J.; Tomas, M.; Menjon, B.; Sunkel, K.; Bau, R., *J. Chem. Soc., Chem. Comm.*, **1984**, 751.
33. Perevalova, E. G.; Reshetova, M. D.; Kokhanyuk, G. M., *Chem. Abstr.*, **1985**, *103*, 37569.
34. Rice, G. W.; Tobias, R. S., *Inorg. Chem.*, **1975**, *14*, 2402.

BIBLIOGRAPHY

Belluco, U., "Organometallic and Coordination Chemistry of Platinum," Academic Press: London, 1974.

Cross R. J.; Mingos, D. M. P., "Organometallic Compounds of Nickel, Palladium, Platinum, Copper, Silver and Gold," Chapman & Hall: London, 1985. Paperback.

Doak, G. O.; Freeman, L. T., "Organometallic Compounds of Arsenic, Antimony and Bismuth," Wiley-Interscience: New York, 1970.

Glockling, F., "The Chemistry of Germanium," Academic Press: London, 1969.

Harrison, P. G., "Organometallic Compounds of Germanium, Tin and Lead," Chapman & Hall: London, 1985. Paperback.

Lesebre, M.; Mazerolles, P.; Satge, J., "The Organic Compounds of Germanium," Wiley-Interscience: London, 1971.

Maitlis, P. M., "The Organic Chemistry of Palladium," Academic Press: New York, 1971.

McKillop, M.; Smith, J. D.; Worrall, I. J., "Organometallic Compounds of Aluminum, Gallium, Indium and Thallium," Chapman & Hall: London, 1985. Paperback.

Poller, R. C., "The Chemistry of Organotin Compounds," Academic Press: New York, 1970.

Puddephat, R. J., "The Chemistry of Gold," Elsevier: Amsterdam, 1978.

Sawyer, A. W., ed., "Organotin Compounds," Marcel Dekker: New York, 1971. 3 vols.

Shapiro H.; Frey, F. W., "The Organic Compounds of Lead," Wiley-Interscience: New York, 1968.

Wardell, J. L., "Organometallic Compounds of Zinc, Cadmium and Mercury," Chapman & Hall: London, 1985. Paperback.

Zuckerman, J. J., ed., "Organotin Compounds: New Chemistry and Applications," American Chemical Society: Washington (D.C.), 1976.

Chapter 7

Metal–Carbon Sigma Bonds. III: The Metalloids

7.1 INTRODUCTION

This chapter will cover the organo chemistry of the elements boron, silicon, germanium, phosphorus, arsenic, selenium and tellurium, along with selected organo derivatives of higher-valent bromine and iodine. Some of these elements would be considered by chemists as "nonmetals" rather than "metalloids." Nevertheless, their organo chemistry dovetails nicely with the elements of the preceding chapter. The total number of reported organo derivatives of these elements probably exceeds that for true organometals, and certainly runs into the tens of thousands. As a result, only the most salient points will receive attention in this discussion.

The following characteristics are general for these organometalloids:

a. *Metalloid–carbon bond thermal stability* — This group of compounds contains the most thermally stable organo derivatives known. Some are stable at 500°C, and even at 1000°C. Such high thermal stability has spurred much of the commercial research on these compounds.

b. *Strong metalloid–oxygen linkages* — The bond between each of these elements and oxygen is thermodynamically very strong, providing a driving force for the formation of such bonds. Despite this, however, many compounds in this category show kinetic resistance towards reaction with air and are perfectly safe to handle.

c. *Extensive catenation* — These elements show a strong tendency to form long chains, ring and cage structures containing metalloid–metalloid bonds. Such bonds become less numerous and more unstable as the Periodic Group number increases.

d. *Extensive multiple bonding* — Elements in this category form multiple bonds to each other and to carbon, nitrogen or oxygen. This ability increases as the Periodic Group number increases.

e. *Importance of attached inorganic groups* — Much of the chemistry of the organometalloids deals with inorganic functional groups attached to the metalloid atom. Metal–carbon bonds are less reactive than for previously discussed organometals, and "perorgano" derivatives are relatively much less important in organometalloidal chemistry than in corresponding chemistry of true metals.

7.2 OXYGEN DERIVATIVES OF THE ORGANOMETALLOIDS

Oxy compounds of organometalloids will have forms such as R—E—OH, R—E—O—X (where X might be a metal, a metalloid, carbon, etc.), and R—E=O.

Hydroxyl groups attached to metalloids (as with the same group on carbon or other nonmetals) are acidic, and such compounds are frequently named as acids: e.g. $CH_3B(OH)_2$, methylboronic acid. Numerous salts and esters of such compounds are known. In some cases, they have a tendency to undergo condensation:

$$2\ R_nE—OH\ \longrightarrow\ R_nE—O—ER_n + H_2O \qquad (7.1)$$

The ease of condensation varies enormously with the nature of E, the type of organic group, the value of n, etc. If there is more than one hydroxyl group on the metalloid, condensation becomes easier, often leading to polymerization:

$$x\ R_nE(OH)_2\ \longrightarrow\ 1/x\ (R_nEO)_x + x\ H_2O \qquad (7.2)$$

Despite this, however, some compounds containing two or even three hydroxides on one atom have been isolated. Organometalloid hydroxides can form by reaction of the corresponding halides with water—a reaction which is often so vigorous that steam is produced. The intermolecular condensation represented in equation 7.2 is characteristic of silicon, and, to a lesser extent, of boron and germanium. These elements do not form stable double bonds with oxygen under ordinary conditions; polymerization occurs instead. Such polymers, especially those of silicon, are stabilized by a form of pi-bonding shown in Figure 7.1. This type of bonding is occasionally termed "dative pi-bonding," and may be considered as a form of Lewis acid–base interaction, in which electrons in p-orbitals on oxygen interact with a d-orbital on silicon, increasing the bond order. This type of bonding is a major reason for the high thermal stability of the Si—O linkage in silicones.

Hydroxides of elements of Groups VA through VIIA can also undergo intramolecular condensation to form metalloid–oxygen double bonds:

$$R_nE(OH)_2\ \longrightarrow\ R_nE=O + H_2O \qquad (7.3)$$

$$R_nE(OH)_3\ \longrightarrow\ \underset{\underset{OH}{|}}{R_nE}=O + H_2O \qquad (7.4)$$

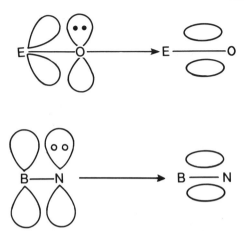

Figure 7.1. Pi-bonding in metalloid–nonmetal systems.

In this respect, they resemble carbon. However, the metalloid–oxygen double bonds have a greater polarity, and in the resonance

$$R_nE{=}O \longleftrightarrow R_n\overset{+}{E}{-}\overset{-}{O}$$

the dipolar single-bonded form predominates. Polymers of these elements are much less common and less stable than those of silicon, boron or germanium.

7.3 MULTIPLY-BONDED ORGANOMETALLOIDS

7.3.1 Compounds with Metalloid–Nonmetal Bonds

The most important class of compounds containing metalloid–nonmetal double bonds are the phosphinemethylenes, $R_3P{=}CH_2$. The P—C linkage is polar, and this polarity makes the linkage reactive. Such compounds were first prepared by Georg Wittig,[1,2] who received the 1970 Nobel Prize in Chemistry for this work. They are frequently termed the "Wittig Reagent" and have wide use in organic synthesis, as for example in the preparation of olefins:

$$
\overset{\displaystyle O}{\underset{\displaystyle \|}{R{-}C{-}R'}} + R_3''P{=}CH_2 \longrightarrow RR'C{=}CH_2 + R_3''PO \qquad (7.5)
$$

These compounds are part of a larger class of species termed "ylids." Such species are characterized by having opposite charges on adjacent atoms covalently bonded to each other.[3] Ylids containing carbon–metalloid bonds are also known for arsenic,[4] antimony and selenium.

The phosphinemethylenes and their arsenic analogs are derivatives of P(V) and As(V) respectively. Compounds containing trivalent P or As forming double bonds to carbon are also known. The heterobenzenes (shown in Figure 7.2) are analogs of pyridine.[5] Spectroscopic investigations indicate that these compounds, along with the Sb and Bi counterparts mentioned in the preceding chapter, show aromatic character. Unlike pyridine, however, phosphabenzene and arsabenzene show very little basic character at the heteroatom. The molecule 2,4,6-triphenylphosphabenzene does form a phosphonium ion (shown in Figure 7.2) with $AlCl_4^-$ as the anion.[6] A stable P=C bonded species is C_6H_5P=$C(NRsi)P(C_6H_5)si$, where si represents the $(CH_3)_3Si$-group.[7]

The most extreme example of multiple bonding in an organophosphorus compound occurs in methinophosphide(methylidenephosphine).[8] This molecule has the formula H—C≡P, is a gas at room temperature, and polymerizes very readily. It undergoes an addition reaction with gaseous hydrogen chloride:

$$HCP + 2\ HCl \longrightarrow H_3CPCl_2 \qquad (7.6)$$

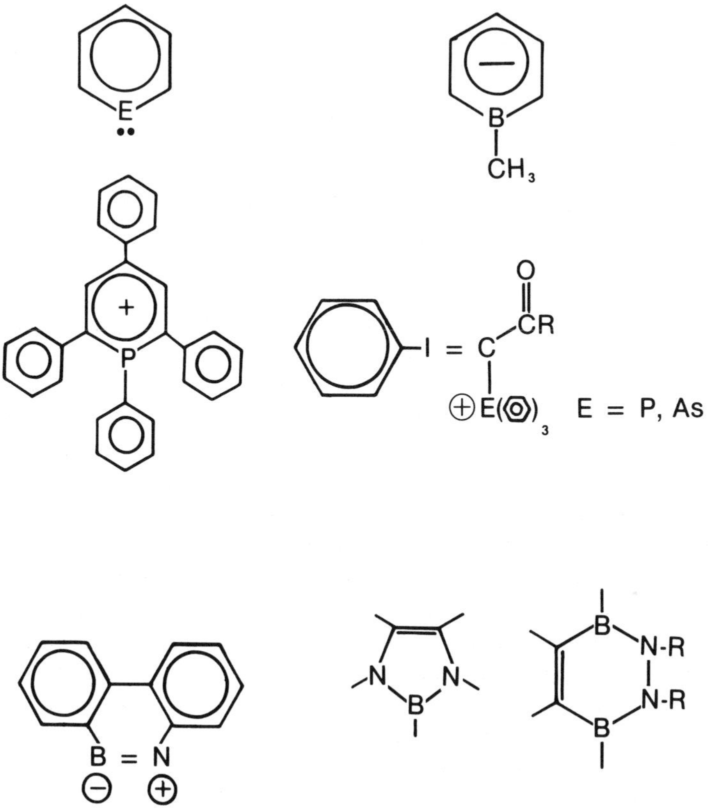

$$E = P, As$$

Figure 7.2. Some examples of metalloid–nonmetal double bonds.

The polymerization of HCP is not unique. Most attempts to prepare doubly bonded metalloid species give polymeric products. Those molecules that have been isolated contain bulky groups that prevent polymerization by causing steric hindrance. One example of a stable Si=C system is the molecule $(CH_3)_2Si=C[Si(CH_3)_3]Si(C_4H_9-t)_2CH_3$, which is a crystalline solid stable enough to have its structure determined.[9] An interesting molecule, one of the few organoarsenicals containing arsenic in two different oxidation states,[10] is $(C_6H_5)_2CH_3As=CHAs(C_6H_5)_2$. The unstable species, $C_6H_5Si\equiv N$, the silicon analog of an isocyanide, has been isolated in an argon matrix at 15K; upon warming, it polymerizes.[11]

Cyclic compounds containing boron–carbon double bonds are known;[12] one example is shown in Figure 7.2. The boron–nitrogen linkage, isoelectronic with carbon–carbon linkages, has received considerable research attention. Various organoboranes containing the —B=N— system have been prepared; some of these are shown in Figure 7.2. The biphenyl analog $C_5H_5\overset{+}{B}—NC_5H_5$ has the two rings twisted at an angle of 43°, but the compound is nonetheless aromatic in character.[13]

A few compounds with C=Se or C=Te bonds are known, such as the ketone analogs, RR'C=Se(Te). These are highly reactive materials, unstable towards air and readily polymerizing to cyclic species.[14,15] Iodonium ylids contain a certain degree of I=C character.[16] The structure shown in Figure 7.2 is an example of an unusual mixed ylid, in which the positive charge is shared between the iodine and the phosphorus (arsenic) atoms.[17]

7.3.2 Compounds with Metalloid–Metalloid Double Bonds

In the early part of this century, workers investigating Salvarsan and related compounds considered them as analogs of azobenzene, writing them as monomers, eg, $C_6H_5As=AsC_6H_5$. This representation persisted for many years, even though physical evidence soon appeared to indicate that they were actually polymeric. Arsenobenzene is now known to be a cyclic polymer. Other efforts to produce multiply bonded metalloid–metalloid compounds also resulted in polymeric compounds; efforts in this area for germanium have been reviewed.[18] The first example of such a compound, when it did appear, released a torrent of similar reports. Within a few years numerous examples of stable compounds containing metalloid–metalloid double bonds have been reported. Table 7.1 lists a number of these.[19–24] The use of bulky ligands seems necessary to prevent polymerization to catenated species—hence the appearance of the 2,4,6-trimethylphenyl(mesitylene) and 2,4,6-tri-*tert*-butylphenyl groups in many of the reported examples. The aromatic nature of the phenyl group may also contribute to the stability of the double bond.[19]

The reported compounds are almost invariably solids, usually crystalline, with melting points in excess of 100°C. Most of them are stable at their melting points, in contrast to analogous species with less bulky groups which

Table 7.1. Selected Examples of Multiply-bonded Organometalloids

Compound	Melting point	Color	Reference
$(Mes)_2Si=Si(Mes)_2$	176°	Orange-yellow	19
$(Dep)_2Ge=Ge(Dep)_2$	215–7°	Yellow	20
$(TMSM)_1P=P (TMSM)_1$	152°, d.	Yellow	21
$(TTBP)As=As(TTBP)$	110–3°	Orange-yellow	22
$(Mes)_2Si=P(TTBP)$	not isolated		23
$(TTBP)_2Ge=P(TTBP)$	155–60°	Orange	24
$(TTBP)As=PCH[Si(CH_3)_3]_2$	118–20°	Orange	22

Abbreviations: *Mes:* mesityl(2,4,6-trimethylphenyl-); *Dep:* 2,6-diethylphenyl-; *TMSM:* tris(trimethylsilyl)-methyl-; *TTBP:* 2,4,6-tris(*tert*-butyl)phenyl-.

polymerize too fast to be isolated at ambient temperatures. They are also colored compounds for the most part. Their chemical reactions strongly resemble those of olefins, especially addition reactions, and proceed rapidly under ordinary conditions. For the most part, they are quite sensitive to air, and must be handled accordingly. Structural studies have been reported for some of these compounds. The disilenes' structures vary somewhat according to the substituents:[19] *trans*-bis(*tert*-butyl)dimesityldisilene has a planar arrangement of the Si_2C_4 framework, while tetramesityldisilene has each SiC_2 group twisted by 18°. The Si—Si bond length is about 20 pm less than the typical Si—Si single bond length. In the digermene,[25] $[\{(CH_3)_3Si\}_2CH]_4Ge_2$, the bond is shortened by about 10 pm, and the angle of twisting increases to 32°. Corresponding values are 23 pm and 26° for the digermene shown in Table 7.1. *Cis-trans* interconversion in disilenes occurs considerably more readily than in olefins,[19] and this increased lability seems to be true for the other multiply-bonded organometalloids.

7.4 CATENATED ORGANOMETALLOIDAL COMPOUNDS

In recent year chemists have uncovered a very extensive group of compounds having metalloid–metalloid single bonds. The reactivity of these compounds depends on the functional groups attached to the metalloids, but the organo derivatives are considerably more stable than hydrogen derivatives. Straight-chain, ring, and cage systems are all known. For the metalloids of Groups V and VI, catenation is known only for lower oxidation states.

Organoboranes having B—B bonds virtually exist exclusively as cage compounds; these species will be discussed in a later section. A few diboranes of formula $R_2B_2(XR'_n)_2$ ($X = 0$, $n = 1$, $X = N$, $n = 2$) have been isolated.[26]

Based on early reports about the silanes, Si_nH_{2n+2}, chemists believed that only short chain silicon compounds were stable. Investigation of organo-polysilanes has shown this to be incorrect; in recent years, an extensive body of data on these compounds has accumulated. An earlier review[27] listed the linear decasilane, $(CH_3)_{22}Si_{10}$, as the largest characterized poly-silane, but more recently, the members of the series from $(CH_3)_{10}Si_5$ to $(CH_3)_{38}Si_{19}$ and $(CH_3)_{48}Si_{24}$ (cyclic compounds all) have been character-ized,[28] and higher members reported.[29] Cage compounds are also known, such as $(CH_3)_{14}Si_8$ and the "bow tie" compound,[30] $(CH_3)_8Si_5$, both shown in Figure 7.3. There seems to be no upper limit to such compounds, and certain ones have become important in polymer chemistry (discussed in the next section). Catenated organogermanes are more limited. Deca-phenyltetragermane is known,[27] and the cyclic series $[(C_6H_5)_2Ge]_x$ (x = 4,5,6) has been isolated and studied,[31] as well as the corresponding methyl analogs. Mixed cyclic organosilagermanes, $(CH_3)_{2n}Si_xGe_{n-x}$ (n = 5,6,7), have also been prepared.[32]

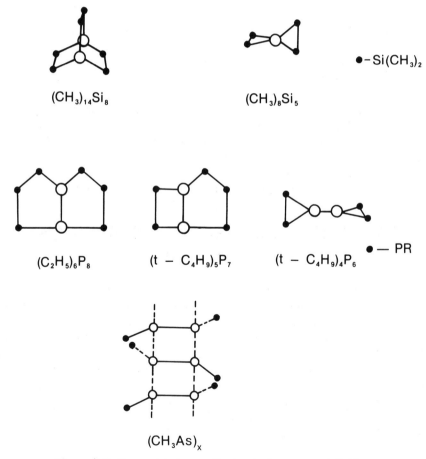

$(CH_3)_{14}Si_8$ $(CH_3)_8Si_5$ $\bullet - Si(CH_3)_2$

$(C_2H_5)_6P_8$ $(t - C_4H_9)_5P_7$ $(t - C_4H_9)_4P_6$ $\bullet - PR$

$(CH_3As)_x$

Figure 7.3. Some structures of catenated organometalloids.

Various tetraorganodiphosphines and -arsines have been reported, but very few higher linear molecules are known. The metalloid–metalloid bond in such molecules is quite labile; in fact, this lability was the reason that $(CH_3)_4As_2$, one of the earliest organometalloids reported, received the common name of "cacodyl" (cf. page 7). Cyclic compounds are more stable and better known.[33,34] These have general formulas $(RE)_n$, with n typically being four to six. These are often colored; pentamethylcyclopentaarsine (earlier known as arsenomethane) exists as a yellow, low-melting (12°C) solid.[34] A high-polymer form of arsenomethane is colored purple and has the "ladder" structure shown in Figure 7.3. The bonds between arsenic atoms in different planes represent "half electron" bonds,[34] and this molecule is of interest for its semiconducting properties. Cage organophosphines have been reported;[35–37] some structures are shown in Figure 7.3.

No cyclic organoselenium or organotellurium compounds containing only the chalconide element in the ring have been reported. Various organo-diselenides, R_2Se_2 and a few diorganopolyselenides, R_2Se_x ($x \geq 3$) are known,[38] as well as dimethyl- and diphenylditelluride.[15] No organoiodine compounds containing iodine–iodine bonds have been reported, although the known stability of species such as I_3^- makes such compounds at least a possibility.

7.5 ORGANOMETALLIC POLYMERS

In recent years polymers using inorganic elements (as opposed to carbon) as the constituent of the "backbone" have received considerable attention from polymer chemists.[39,40] Originally, these materials were important because of their high thermal stability; more recently, many have been recognized as electric conductors or semiconductors. The most investigated and commercially the most important of these polymers are the silicones. These compounds have silicon–oxygen backbones, with organic groups (usually methyl) bonded to the silicon atoms. The most common silicones have the general formula $[(CH_3)_2SiO]_x$, where x can range from 3 on up to many thousands. These are prepared in the following manner:

$$CH_3Cl + Si \xrightarrow{Cu} (CH_3)_2SiCl_2 \qquad (7.7)$$

$$(CH_3)_2SiCl_2 + 2\,OH^- \longrightarrow [(CH_3)_2SiO]_4 + 2\,Cl^- + H_2O \qquad (7.8)$$

$$\frac{x}{4}(CH_3)_8Si_4O_4 \xrightarrow{OH^-} [(CH_3)_2SiO]_x \qquad (7.9)$$

The cyclic tetramer, octamethylcyclotetrasiloxane, is a volatile liquid which may be removed by distillation. The degree of polymerization may be controlled, and the products may be liquids or solids. Addition of trimethylchlorosilane in equation 7.8 provides "end groups" and gives linear molecules such as $(CH_3)_{10}Si_4O_4$, while the addition of methyltrichlorosilane or $SiCl_4$ enables cross-linking, yielding rigid three-dimensional solids. Other chlorosilanes, such as $CH_3(C_6H_5)SiCl_2$ might be used if unusual properties in the products were needed.

Silicones combine the best features of glass and of organic polymers. The properties that make them so useful are:

a. *Thermal stability* — Silicone polymers remain undecomposed at temperatures up to 1000°C and even higher. In addition, they are insulating materials, which have virtually no thermal conductivity.
b. *Chemical inertness* — Silicone polymers are among the most inert chemical materials known. They show no tendency to be oxidized by air, even after years of exposure, nor do they react with most chemical reagents. Biochemical processes do not affect them.
c. *Water impermeability* — Silicone polymers do not react with water, dissolve in water, absorb or even adsorb water. They are completely water repellant.
d. *Low surface tension* — Fluid silicone polymers have very low surface tension, and lower the surface tension of other liquids, such as water.

These various properties, along with the wide range of liquid or solid forms in which they might be prepared, give silicones an extensive array of commercial applications, only a few of which can be mentioned here. Silicones can be used to disperse foam in standing bodies of water (or in human digestive systems). Silicone-based fluids can be sprayed onto adsorbant surfaces (cement, mortar, cloth, leather, paper, etc.) to protect them from water. Silicone rubbers and plastics have been used for various purposes in surgery and the health sciences, primarily for their resistance to biological attack. The toy "silly putty" owes its attraction to the unique properties of silicones.

Other classes of organometallic polymers have been less fully developed, although they are becoming increasingly important. Organophosphonitriles, $(R_2PN)_x$, also called poly(organophosphazenes), are currently being investigated for a variety of purposes.[40] Polymers containing carborane units in combination with silicone units show marked thermal stability and have been proposed for use as partitioning phases in chromatography.[41] High molecular weight polysilanes have been reported.[29,40,42] Phenylmethylpolysilanes, with molecular weights of about 300,000 (corresponding to 2500 units or more), shows semiconducting properties.[42] The corresponding dimethylpolysilanes, upon heating to 400°, rearranges to give a

$$
\text{polymer with the structure } (-\underset{\underset{CH_3}{|}}{\overset{\overset{H}{|}}{Si}}-CH_2-)_x, \text{ which in turn, upon heating}
$$

to 900°, gives β-silicon carbide having fibrous properties.[40] Similarly, polysilazanes, $(R_2SiNH)_x$, can be used to prepare silicon nitride.[40] Polymethylpolyarsine (purple arsenomethane) also shows semiconducting properties. The expanding interest in semiconductors and ceramics indicate that this area should grow substantially in years to come.

Inorganic and organic polymers containing organometal(loid)s as side groups are also known and have some commercial interest. The most important of these are the antifouling paints containing tri-n-butyltin groups; these compounds will be discussed in Chapter 12.

7.6 ORGANOBORON COMPOUNDS

Unlike their aluminum analogs, simple alkyl- or arylboranes show no tendency to polymerize through three-center, two-electron bonds. Compounds with boron-hydrogen bonds (R_2BH, RBH_2) do exist as dimers and may be considered as organo derivatives of B_2H_6. The simple organoboranes are air-sensitive, and react with oxygen at varying rates, depending on the organic groups present. Triphenylborane is a good Lewis acid: it can accept an electron to form an anion radical (see Chapter 3.4.2), combine with neutral donor molecules, and accept anions, particularly carbanions. Tetraphenylborate salts, such as Na^+ $(C_6H_5)_4B^-$, and other tetraarylborates have a number of uses in analytical chemistry, including the quantitative determination of K^+. The salt $(i\text{-}C_5H_{11})_4N^+$ $(i\text{-}C_5H_{11})_4B^-$, in which the two ions are isoelectronic and virtually identical in size, has been used as an electrolyte standard. Tetraphenylborate ion dissolves in hydrocarbons as well as in water, making it useful for cell membrane investigations (see Chapter 11).

Organoboron chemistry has been strongly influenced by two major factors. The first was the hydroboration reaction, which has become a major synthetic tool in organic synthesis.[43,44] One such use is in the isomerization of olefins:

$$2\ CH_3CH{=}CHCH_3 + (R_2BH)_2 \rightleftharpoons 2\ CH_3CH_2CH(BR_2)CH_3 \quad (7.10)$$

$$2\ CH_3CH_2CH(BR_2)CH_3 \rightleftharpoons 2\ CH_3CHCH{=}CH_2 + (R_2BH)_2 \quad (7.11)$$

$$2\ CH_3CHCH{=}CH_2 + (R_2BH)_2 \rightleftharpoons R_2BCH_2CH_2CH_2CH_3 \quad (7.12)$$

The n-butyldialkylborane can be separated by physical techniques and undergo dehydroboration to form 1-butent.

The second major influence on organoborane chemistry was the discovery of the carboranes. These species are neutral molecules or ions, usually of formula $B_nC_2H_{n+2}$, in which the carbon atoms have been directly incorporated into the boron cage. The first and best known of the carboranes is 1,2-dicarbaclosododecaborane-12, shown in Figure 7.4. It was prepared in the following way:

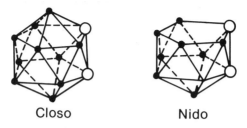

Closo   Nido

0 — Phenylenebis (dimethylarsine)

Ylid-stabilized Organometals(M = Cu, Ag, Au)

Periodinane
Figure 7.4. Various organometalloids.

$$B_{10}H_{14} + 2\ CH_3CN \longrightarrow B_{10}H_{12}(NCCH_3)_2 + H_2 \qquad (7.13)$$

$$B_{10}H_{12}(NCCH_3)_2 + HC\equiv CH \longrightarrow$$
$$B_{10}H_{10}C_2H_2 + H_2 + 2\ CH_3CN \quad (7.14)$$

The intermediate adduct is usually not isolated. Other Lewis bases, such as dimethylsulfide, have also been used. The compound $1,2\text{-}B_{10}C_2H_{12}$ is often given the trivial name "o-carborane." It is a crystalline solid melting at 285–287°C. When heated to 485°C, this compound rearranges to the

1,7 isomer ("m-carborane"), which, upon further heating to 620°C, forms the 1,12-isomer ("p-carborane"). Such high thermal stability has spurred work on carboranes and their ability to be incorporated into polymers.

The two hydrogens on the carbon atoms of o-carborane are different in their properties than the hydrogens on the boron atoms; they will undergo lithiation when treated with n-butyllithium to give $Li_2C_2B_{10}H_{10}$. This in turn will react with a wide variety of reagents, enabling the carborane moiety to be incorporated into larger molecules or polymerized:

$$LiC\underset{B_{10}H_{10}}{\diagdown O \diagup}CLi + Cl_2 \longrightarrow ClC\underset{B_{10}H_{10}}{\diagdown O \diagup}CCl \tag{7.15}$$

$$LiC\underset{B_{10}H_{10}}{\diagdown O \diagup}CLi + ClC\underset{B_{10}H_{10}}{\diagdown O \diagup}CCl \longrightarrow \frac{2}{x}\left[-C\underset{B_{10}H_{10}}{\diagdown O \diagup}C-\right]_x \tag{7.16}$$

The hydrogens on boron can also be replaced if necessary, but such derivatives are less widely known than the carbon-substituted species.

Various other neutral carboranes of formula $B_nC_2H_{n+2}$ have been reported.[45–47] These are the *closo* compounds—the boron atoms and the carbon atoms form a completely closed polyhedron involving delocalized pi-electrons. All hydrogens are terminal (as opposed to bridging). The compound $1,2-C_2B_4H_6$ would have an octahedral structure with the two carbon atoms at adjacent positions and would be named 1,2-dicarbaclosohexaborane-6. In the strictest sense, the term "carborane" applies only to such species. However, it is often extended to include other boron–carbon compounds that have briding hydrogens. Loss of one BH unit from a *closo* structure yields a *nido* structure (shown in Figure 7.4) having an open face; one example of a *nido* compound is $B_9C_2H_{13}$. Removal of a BH unit from a *nido* structure gives an *arachno* structure.[48]

Carboranes can form numerous compounds with metals and metalloids. Some of these involve incorporation of a nonmetal or metalloid (eg, P, As, S) into the boron cage structure.[49] More often, the carborane might lose a BH group to form an ion ("carbollide"), which then reacts with a transition metal to form compounds similar to the aromatic synergistic complexes discussed in Chapter 9. These compounds are called "metallocarboranes" and are becoming increasingly common.[46–49]

7.7 ORGANOSILICON AND ORGANOGERMANIUM COMPOUNDS

Certain aspects of the very extensive body of organosilicon chemistry have already been discussed. Much of this has come into being due to the great commercial interest in the silicones and their chemical precursors. Germanium, being a much scarcer element than silicon, has not had its organic chemistry developed to the same extent.

Both elements belong to Group IVA—the same group as carbon. Much early work on organosilicon chemistry began as an attempt to prepare silicon analogs of organic compounds. However, the differences between the two elements are numerous and greatly outweigh any similarities. In general, where carbon forms double or triple bonds with itself or other elements, silicon will form polymers containing single bonds only (exceptions to this were discussed in Chapter 7.3).

These elements can form organo compounds having one, two, three or four element carbon bonds. The tetraorgano species, R_4E, are covalent materials having low melting and boiling points and resembling the corresponding hydrocarbons in their physical properties. They are chemically perhaps the least reactive of all organometalloids, being resistant to bond cleavage under most ordinary conditions. As are all organometal(loid)s, these compounds are thermodynamically unstable towards reaction with oxygen; however, they require activation energy to begin reaction. Tetramethylsilane, $(CH_3)_4Si$, has been used for many years as the standard reference compound for measuring proton chemical shifts in proton nuclear magnetic resonance spectroscopy.

The most reactive of the common organosilicon and -germanium compounds are the halides. The Si—X and Ge—X bonds are labile, especially towards protonic solvents. This makes them very useful as starting materials for synthesis; it also means they must be handled carefully and protected from moisture. Reaction with water gives siloxanes or germoxanes (see equation 7.1), but the rate and extent of reaction depend very much on the other groups attached. Most silanols (having Si—OH linkages) condense readily, but triphenylsilanol is readily isolated and may be stored. The hydrolysis reaction may be used to generate acids as reagents. Hydrazoic acid, HN_3, is unstable, poisonous and potentially explosive, making it difficult and dangerous to use as a reagent. The trimethylsilyl derivative, $(CH_3)SiN_3$, is much easier to handle and reacts with water

$$2\ (CH_3)_3SiN_3 + H_2O \longrightarrow (CH_3)_3SiOSi(CH_3)_3 + 2\ HN_3 \quad (7.17)$$

thereby generating hydrazoic acid in situ. Most other anionic species attached to silicon are hydrolyzed in the same way. The corresponding germanium compounds hydrolyze more slowly under comparable conditions, and the reaction may be reversed in strong acids:

$$R_3GeOH + HBr \longrightarrow R_3GeBr + H_2O \quad (7.18)$$

The ready reaction of chlorotrimethylsilane with protonic materials, and the subsequent ability of the bonds to undergo hydrolysis, has made this compound and related ones very useful for chromatographic separations of organic compounds having —OH, —NH$_2$, or —NH— groups. When treated with $(CH_3)_3SiCl$, these groups form Si—N or Si—O bonds, removing the hydrogen as HCl and destroying any intermolecular hydrogen bonding. The resulting derivatives are more volatile than the starting materials, and may readily be separated by vapor phase chromatography. If desired, they may then be hydrolyzed to reform the starting materials. This form of silylation has become extensively used in biochemistry for the separation and determination of complex mixtures.

Unlike carbon, both silicon and germanium can achieve coordination numbers of five or six, although usually only with the most electronegative elements. Some complexes of organosilicon compounds are known, such as $(C_6H_5)_3Si(bipy)^+I^-$, but they are relatively uncommon. Germanium forms a larger number of complexes, but considerably less than has been reported for tin. Organogermanium–oxygen compounds do not form high polymers (there are no germanium analogs of the silicones), and the oligomers that do form are sensitive towards water and acids.

7.8 ORGANOPHOSPHORUS AND ORGANOARSENIC COMPOUNDS

Phosphorus and arsenic differ from the preceding metalloids in that they have extensive organo chemistries involving two different, readily interconverted oxidation states; in this, they resemble their heavier congenor antimony. Both elements have had widespread use in biological applications; in addition, organophosphorus compounds, particularly the Wittig Reagent, have many uses in organic syntheses.

Organophosphorus(III) and organoarsenic(III) compounds all have a lone pair of electrons on the metalloid atom, making them potentially donor molecules. Triorgano compounds, in fact, have been widely used in transition metal chemistry to form donor-acceptor complexes, and many thousands of such species have been reported. Chelating agents using two metalloid atoms are also known; a very familiar one is o-phenylenebis(dimethylarsine) ("diars"), shown in Figure 7.4. The lone pair can react with organic halides to give oxidative addition:

$$R_3E: + R'X \longrightarrow [R_3ER']^{\oplus} + X^- \qquad (7.19)$$

Similarly, whenever a hydroxyl group is attached to an organophosphorus (III) compound, the hydrogen attached to oxygen migrates to phosphorus:

$$R_2\ddot{P}—OH \longrightarrow R_2P(:O)H \qquad (7.20)$$

This is closely related to the Arbusov rearrangement,

$$(RO)_3P: + R'X \longrightarrow (RO)_2P(:O)R' + RX \qquad (7.21)$$

which is postulated as passing through a phosphonium intermediate. Corresponding rearrangements are not known for arsenic.

Phosphonium salts can also be used to prepare phosphorus ylids (see Chapter 7.3.1) in the following manner:

$$(CH_3)_4P^+ \; X^- + C_6H_5Li \longrightarrow (CH_3)_3P{=}CH_2 + C_6H_6 + LiX \quad (7.22)$$

The corresponding ylids of other metalloids may be prepared in a similar fashion. The presence of phosphorus can add stability to the ylid carbon in the latter's compounds, and has enabled the isolation of organo derivatives of transition metals such as copper, silver or gold.[50] The structure of one such compound is shown in Figure 7.4.

Pentaorgano compounds of phosphorus and arsenic are known almost exclusively for aryl derivatives [eg, $(C_6H_5)_5P$], although pentamethylarsorane, $(CH_3)_5As$, has been reported.[51] These compounds have a trigonal bipyramidal structure. One metalloid–carbon bond undergoes ready cleavage to form the corresponding onium salts, but, unlike the antimony analog, pentaphenylphosphorane or -arsorane do not add a sixth phenyl group to form $(C_6H_5)_6E^-$, presumably for steric reasons.

Formation of phosphorus–carbon and arsenic–carbon bonds will occur in a variety of natural organisms, usually as a mechanism of detoxification. This subject will be discussed in Chapter 13.

7.9 ORGANOSELENIUM AND ORGANOTELLURIUM COMPOUNDS

Organoselenium and organotellurium compounds are numerous. However, the organo chemistry of these two elements developed rather independently of either "pure" organic chemistry or organometallic chemistry, and have had less influence than their numbers might suggest. As with phosphorus and arsenic, these elements have organo derivatives in two different oxidation states. Actually, a third state is also possible; however, except for some ill-defined selenones, R_2SeO_2, organoselenium(VI) or organotellurium(VI) compounds have not been isolated, although there seems no reason why a molecule such as $C_6F_5TeF_5$ might not be stable enough for characterization. Organo derivatives of these two elements in both the plus two and plus four states are common, and interconversion between them seems to be more labile than for phosphorus or arsenic.

Tetraorgano derivatives are rather unstable, and only a few have actually been isolated. Tetrakis(trifluoromethyl)tellurane has been reported:[52]

$$(CF_3)_2TeCl_2 + (CF_3)_2Cd \xrightarrow[-10°C]{CH_3CN} (CF_3)_4Te + CdCl_2 \quad (7.23)$$

This compound melts at $-45°C$ and is stable at $0°C$ in the absence of light. In general, efforts to form the tetraalkyls lead to reductive elimination:

$$R_3EX + RLi \longrightarrow LiX + [R_4E] \longrightarrow R_2E + R_2 \quad (7.24)$$

Triorganoselenonium and triorganotelluronium salts are the best known and most stable organo derivatives in the $+4$ oxidation state. Diorganotellurium(IV) oxygen and halogen derivatives tend to be polymeric in the solid state. As with the isoelectronic As and Sb(III) compounds, the lone pair of electrons is stereochemically active. Compounds of formula R_2TeX_2 usually have a trigonal bipyramidal structure, with the lone pair occupying an equatorial position and distorting the geometry. Dimethyltellurium didiodide, $(CH_3)_2TeI_2$ and other dihalides exist in two series: the alpha forms—covalent and monomeric; and the beta forms—ionic dimers of structure $(CH_3)_3Te^+$ $CH_3TeI_4^-$. A compound of reported formula $(CH_3)_2TeI_4$ is actually a 1:1 adduct of the diiodide and molecular iodine.

The most important derivatives of the $+2$ oxidation state are the diorganoselenium or diorganotellurium compounds, often called dialkylselenides(tellurides). These are covalent, volatile species with extremely unpleasant and tenacious odors. As with the corresponding compounds of sulfur, phosphorus and arsenic, the dialkyls of selenium and tellurium can form complexes with metals, although their uses in this respect have not been much investigated. They are readily oxidized by halogens or organic halides.

Organoselenium compounds have an extensive biochemistry, usually as analogs of sulfur; one example is the compound selenomethionine, $CH_3SeCH_2CH_2CH(NH)CO_2H$. Dimethylselenium and dimethyltellurium can be formed by the action of organisms on selenite and tellurite respectively (see Chapter 13).

7.10 ORGANOIODINE COMPOUNDS

The organo chemistry of iodine is usually considered as part of organic chemistry proper, and the organo derivatives of monovalent iodine will not be considered here. However, iodine can form organo compounds in higher oxidation states (often dubbed "hypervalent" compounds). Derivatives of trivalent iodine are numerous, and the aryl compounds seem to be the most stable. Triphenyliodine is a yellow solid that explodes when heated to 25°C, but reacts with triphenylborane:[53]

$$(C_6H_5)_3I + (C_6H_5)_3B \longrightarrow (C_6H_5)_2I^+ B(C_6H_5)_4^- \qquad (7.24)$$

In this respect, triphenyliodine resembles tetraphenyltellurium and pentaphenylbismuth. Numerous diphenyliodonium salts are known,[54] and there are various monophenyliodine(III) compounds, such as iodosobenzene, C_6H_5IO, and phenyliodine diacetate, $C_6H_5I(C_2H_3O_2)_2$. Most of the reports on these compounds in the chemical literature involve their use as oxidizing reagents in organic syntheses.

A variety of "periodinanes" containing I(V) have been reported.[55] These all have a single organic group bonded to iodine, along with four fluorines, four oxygens, or a mixture of fluorines and oxy ligands. The most stable of these compounds is shown in Figure 7.4. It is thermally stable, melting at 205–210°C (with sublimation), but chemically quite reactive, oxidizing aqueous hydrochloric acid to chlorine. Reduction of the periodinane gives the corresponding iodinane—a trivalent iodine species with an iodine-carbon bond and two intramolecular iodine-oxygen bonds.

Organo compounds containing higher-valent chlorine or bromine are very rare. Chloronium or bromonium species of type R_2X^+ have been postulated as reaction intermediates[56] but almost never isolated. A brominane—the bromine analog of the iodinane just mentioned—has been isolated;[57] it is a thermally stable solid, melting at 153–154°.

REFERENCES

1. Vedejs, E., *Science,* **1980,** *207,* 42.
2. Wittig, G., *Science,* **1980,** *210,* 600.
3. Johnson, A. W., "Ylid Chemistry," Academic Press: New York, 1966, pp. 1–5.
4. Huang, Y. Z.; Shen, Y. C., *Adv. Organomet. Chem.,* **1982,** *20,* 115.
5. Ashe, A. J., *Accts. Chem. Res.,* **1978,** *11,* 153.
6. Dave, T. N.; Kaletsch, H.; Dimroth, K., *Angew. Chem.,* **1984,** *96,* 984.
7. Appel, R.; Knoch, F.; Laubach, B.; Sievers, R., *Chem. Ber.,* **1983,** *116,* 1873.
8. Gier, T. E., *J. Am. Chem. Soc.,* **1961,** *83,* 1769.
9. Wiberg, N.; Wagner, G.; Mueller, G., *Angew. Chem.,* **1985,** *97,* 220.
10. Schmidbaur, H.; Nusstein, P., *Organometallics,* **1985,** *2,* 344.
11. Bock, H.; Dammel, R., *Angew. Chem., Int. Ed. Engl.,* **1985,** *24,* 111.
12. Herberich, G. E.; Becker, H. J.; Hessner, B.; Zelenka, L., *J. Organometal. Chem.,* **1985,** *280,* 147.
13. Boese, R.; Finke, N.; Henkelmann, J.; Maier, G.; Paetzold, P.; Reisenauer, H. P.; Schmid, G., *Chem. Ber.,* **1985,** *118,* 1644.
14. Silverman, R. B., in "Organic Selenium Compounds" (Klayman, D. J.; Günther, W. H. H., eds.), J. Wiley & Sons: New York, 1973, pp. 245–261.
15. Bagnall, K. W., "The Chemistry of Selenium, Tellurium and Polonium," Elsevier: Amsterdam, 1966, pp. 174–175.
16. Koser, G. F., "The Chemistry of Functional Groups, Suppl. D," J. Wiley & Sons: New York, 1983, pp. 774–800.
17. Moriarty, R. M.; Prakash, I.; Prakash, O.; Freeman, W. A., *J. Am. Chem. Soc.,* **1984,** *106,* 6082.
18. Satge, J., *Adv. Organomet. Chem.,* **1982,** *21,* 241.
19. West, R., *Science,* **1984,** *225,* 1109.
20. Snow, J. T.; Murakami, S.; Masamune, S.; Williams, D. J., *Tet. Lett.,* **1984,** *25,* 4191.
21. Cowley, A. H.; Kilduff, J. E.; Newman, T. H.; Pakulski, M., *J. Am. Chem. Soc.,* **1982,** *104,* 5820.
22. Cowley, A. H.; Lasch, J. G.; Norman, N. C.; Pakulski, M., *J. Am. Chem. Soc.,* **1983,** *105,* 5506.
23. Smit, C. N.; Lock, F. M.; Bickelhaupt, F., *Tetrahedron Lett.,* **1984,** *25,* 3011.
24. Escudie, J.; Couret, C.; Satge, J.; Andrianarison, Mbolatiana; Andriamizaka, J.-D., *J. Am. Chem. Soc.,* **1985,** *107,* 3378.
25. Hitchcock, P. B.; Lappert, M. F.; Miles, S. J.; Thorne, A. J., *J. Chem. Soc., Chem. Comm.,* **1984,** 480.
26. Coyle, T. D.; Ritter, J. J., *Adv. Organomet. Chem.,* **1972,** *10,* 237.
27. Gilman, H.; Atwell, W. H.; Cartledge, F. K., *Adv. Organomet. Chem.,* **1972,** *10,* 237.
28. Brough, L. F.; West, R., *J. Am. Chem. Soc.,* **1981,** *103,* 3049.

29. West, R., in "Comprehensive Organometallic Chemistry," Pergamon: London, 1982, Volume 2, pp. 365–397.
30. Boudjouk, P.; Sooriyakumaran, S., *J. Chem. Soc., Chem. Comm.*, **1984,** 777.
31. Ross, L.; Draeger, M., *Z. Anorg. Allgem. Chem.*, **1984,** *515,* 141.
32. West, R.; Carberry, E., *Science,* **1975,** *189,* 179.
33. Cullen, W. R., *Adv. Organomet. Chem.*, **1966,** *4,* 145.
34. Smith, L. R.; Mills, J. L., *J. Organometal. Chem.*, **1975,** *84,* 1.
35. Baudler, M.; Makowa, B., *Angew. Chem.*, **1984,** *96,* 976.
36. Baudler, M.; Daerr, E.; Binseh, G.; Stephenson, D. S., *Z. Naturforsch.*, **1984,** *39B,* 1671.
37. Baudler, M.; Michels, M.; Hahn, J.; Pieroth, M., *Angew. Chem.*, **1985,** *97,* 514.
38. Irgolic, K. J.; Kudchadker, M. V., in "Selenium" (Zingaro, R. A.; Cooper, W. C., eds.), Van Nostrand Reinhold: New York, 1974, pp. 408–540.
39. Carraher, C. E.; Sheats, J. E.; Pittman, C. U., eds., "Organometallic Polymers," Academic Press: New York, 1978.
40. Allcock, H. R., *Chemical & Engineering News* (March 18, 1985), 22.
41. Anon., *Chemical & Engineering News* (March 22, 1971), 46.
42. West, R.; David, L. D.; Durovich, P. I.; Stearley, K. L.; Srinivasan, K. S. V.; Yu, H., *J. Am. Chem. Soc.*, **1981,** *103,* 7352.
43. Brown, H. C., *Chemical & Engineering News* (March 5, 1979), 24.
44. Brown, H. C., *Science,* **1980,** *210,* 485.
45. Onak, T., *Adv. Organomet. Chem.*, **1965,** *3,* 263.
46. Grimes, R., "Carboranes," Academic Press: New York, 1970.
47. Dunks, G. B.; Hawthorne, M. F., *Accts. Chem. Res.*, **1973,** *6,* 124.
48. Douglas, B.; McDaniel, D. H., Alexander, J. J., "Concepts and Models in Inorganic Chemistry," 2nd. Ed., J. Wiley & Sons: New York, 1982, pp. 664–684.
49. Todd, L. J., *Adv. Organomet. Chem.*, **1970,** *8,* 87.
50. Schmidbaur, H., *Accts. Chem. Res.*, **1975,** *8,* 62.
51. Mitschke, K. H.; Schmidbaur, H., *Chem. Ber.*, **1973,** *106,* 3645.
52. Naumann, D.; Wilkes, B., *J. Fluor. Chem.*, **1985,** *27,* 115.
53. Wittig, G.; Claus, K., *Annalen.*, **1953,** *578,* 136.
54. Beringer, F. M.; Dehn, J. W.; Winicov, M., *J. Am. Chem. Soc.*, **1960,** *82,* 2948.
55. Nguyen, T. T.; Amey, R. L.; Martin, J. C., *J. Org. Chem.*, **1982,** *47,* 1024.
56. Olah, G., "Halonium Ions," Wiley/Interscience: New York, 1975.
57. Nguyen, T. T.; Martin, J. C., *J. Am. Chem. Soc.*, **1980,** *102,* 7382.

BIBLIOGRAPHY

Arsenic
Raiziss, G. W.; Gavron, J. L., "Organic Arsenical Compounds" Chemical Catalog Company: New York, 1923.
Boron
Steinberg, H., "Organoboron Chemistry" 2 vol., J. Wiley & Sons: New York, 1964.
Germanium
Glockling, F., "The Chemistry of Germanium," Academic Press: New York, 1969.
Rijkens, F.; Van der Kerk, G. J. M., "Organogermanium Chemistry," Germanium Research Committee: Utrecht, Netherlands, 1964.
Phosphorus
Kosolapoff, G. M.; Maier, L., "Organic Phosphorus Compounds," 7 vol., Wiley-Interscience: New York, 1972.
Kirby, A. J.; Warren, S. G., "The Organic Chemistry of Phosphorus," Elsevier: Amsterdam, 1967.
Gefter, E. L., "Organophosphorus Monomers and Polymers" (G. M. Kosolapoff, trans.), Associated Technical Services: Glen Ridge, New Jersey, 1962.
Selenium
Klayman, D. L.; Guenther, W. H. H., eds., "Organic Selenium Compounds: Their Chemistry and Biology," Wiley-Interscience: New York, 1973.

Silicon
Eaborn, C., "Organosilicon Compounds," Butterworths: London, 1960.
Rochow, E. G., "An Introduction to the Chemistry of the Silicones," J. Wiley & Sons: New York, 2nd Ed., 1951.
Post, H. W., "Silicones and Other Organic Silicon Compounds," Reinhold: New York, 1949.
Tellurium
Irgolic, K. J., "The Organic Chemistry of Tellurium," Gordon and Breach: New York, 1974.

Chapter 8

Metal–Carbon Synergistic Bonds. I: Mononuclear Compounds of Divalent Carbon

8.1 INTRODUCTION

The word "synergistic" means "mutually reinforcing." When applied to metal–carbon bonds, it refers to a two-component linkage. The "sigma" component involves the donation of electron density from an organic molecule to a metal atom. The second or "pi" component involves electron density transfer in the opposite direction—from metal to organic ligand. This second component is often termed the "back bond" or a "dative pi-bond."

Formation of such bonds requires the following conditions:

Metal — Must have (a) one or more empty atomic orbitals of appropriate symmetry to accept electrons; and (b) electrons in orbitals of suitable energy and symmetry to be donated;

Organic moiety — Must have (a) electrons available for donation; and (b) empty orbitals (usually *p*-orbitals) of energy and symmetry suitable for acceptance of electrons.

The requirements for the metals limit these compounds to transition metals, predominantly those found in the central portions of the Periodic Table, while the requirements for the organic molecules confines such ligands to divalent carbon species or unsaturated hydrocarbons. Historically, these compounds were among the earliest organometals to be reported (Zeise's salt; nickel tetracarbonyl), but for decades they were overshadowed by the developing chemistry of the sigma-bonded organometal(loid)s, and it was not until the synthesis of ferrocene that the nature of the bonding in these molecules was understood. Nowadays the synergistic organometals receive the lion's share of research attention and have extensive coverage in standard inorganic textbooks.

For the purposes of discussion, this book will subdivide the synergistic organometals as follows: mononuclear (ie, one metal atom) compounds containing divalent carbon (carbon monoxide; :CO; organic isocyanides, RNC:; carbenes, R_2C: and so forth); mononuclear compounds containing unsaturated hydrocarbons; and polynuclear compounds of both types. The first two categories are not clear-cut—there are numerous examples of mixed species, such as benzenechromium tricarbonyl, $C_6H_6Cr(CO)_3$—but will suffice for an introduction.

8.2 SYNERGISTIC METAL–CARBON(II) BONDING

Divalent carbon species all have a lone pair of electrons on the carbon atom, which can be donated to a metal atom. Cyanide ion, $:CN^-$, which might well be included in this category, forms a wide range of stable complexes with transition metals. The donation from carbon monoxide and other neutral derivatives of divalent carbon is considerably weaker than from cyanide ion, and most of the complexes owe their stability to the pi-component. Figure 8.1 shows the orbital interactions that form the metal–carbon bond in a carbonyl compound. The carbon–oxygen pi antibonding orbital receives the electrons from the metal. Detailed discussion of the theory behind this type of bond appears elsewhere.[1-4]

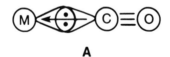

A

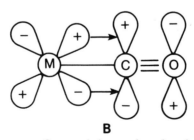

B

Figure 8.1. Orbital diagram of synergistic metal–carbon bond in metal carbonyls: (A) Donation from carbon to metal; (B) Donation from metal to carbon.

The extent of "backbonding" and the effect of other ligands on this phenomenon have interested many researchers. Since electron density is being transferred from metal to a ligand antibonding orbital, bonding theory predicts that this should weaken the energy of the bond in question. In carbon monoxide the carbon—oxygen stretching band has an intense absorbance at 2148 cm^{-1}. When carbon monoxide bonds to a metal atom, the resulting "backbond" weakens the carbon—oxygen bond, causing this absorbance to shift to a lower value; the extent of the shift is an indication of the degree to which this "back-bonding" occurs. The same is true, although less spectacularly, for other divalent carbon species and also for olefins, making vibrational spectroscopy a valuable technique for investigating such systems.[5]

As more and more compounds containing metal—carbon(II) species were reported, stoichiometric patterns appeared. These became explicable by a formalism known as the Effective Atomic Number (E.A.N.) Rule, also known as the Rule of 18. Originally developed by Sidgwick to explain the stoichiometry of covalent compounds,[6] it is now used almost exclusively for synergistic organometals and related compounds. The rule might be stated as follows:

The stable ligand configuration around a metal atom will be that in which the metal has the effective atomic number of the next highest noble gas (in which the metal has eighteen electrons in its outermost shell).

Although there are occasional exceptions, the great majority of metal carbonyls and other transition metal complexes follow this rule, making it a useful tool for prediction and explanation. In applying this rule, the following electron count is used for ligand donation to each metal atom:

terminal CO groups	2 electrons each
bridging CO groups	1 electron each
σ-bonded ligands	1 electron each
metal—metal bonds	1 electron each
unsaturated hydrocarbons	2 electrons per π bond

To illustrate this, consider the very simple molecule nickel tetracarbonyl, $Ni(CO)_4$. The nickel atom is considered to be in the oxidation state of zero, giving it 10 electrons ($3d^8 4s^2$). Each of the four carbonyls is terminal (ie., bonded to only one metal atom), and donates two electrons. The four groups donate a total of eight electrons, which, added to the 10 outer shell electrons of nickel, give a total of 18. Alternatively, an atom of nickel has 28 electrons (the same as its atomic number); add to this the eight electrons from the four carbonyl groups, and the total is 36, the same as the atomic number of krypton, the next higher noble gas in the Periodic Table. Another example is methylmanganese pentacarbonyl, $CH_3Mn(CO)_5$. Manganese has atomic number 25; add to this the 10 electrons from the five carbonyl groups and one electron from the sigma-bonded methyl group, and the total comes to 36, the E.A.N. of krypton. Other examples of the E.A.N. Rule, including its use for unsaturated hydrocarbon and polynuclear complexes, will be given in this chapter and the next two.

8.3 NEUTRAL METAL CARBONYLS

Comparatively few neutral molecules of formula $M_1(CO)_n$ have been reported. These are shown in Table 8.1. They have been the subject of extensive spectroscopic and structural investigations, particularly iron pentacarbonyl.[7,8] They show the expected geometries: hexacarbonyls are octahedral, pentacarbonyls are trigonal bipyramidal, and tetracarbonyls are tetrahedral. Analyses of their vibrational spectra revealed that the M—C—O triatomic skeleton is linear. Carbon–oxygen stretching vibrations occur between 2000 and 2050 cm^{-1}; several intense bands are usually seen in this region. As previously noted, the lower frequency relative to free carbon monoxide (2148 cm^{-1}) has been attributed to the transfer of electron density from the metal atom into the antibonding orbital of carbon monoxide.

The carbonyls shown in Table 8.1 obey the E.A.N. Rule, except for $V(CO)_6$, which has only 17 electrons. Analogy with the corresponding carbonyls of the metals chromium through nickel would predict that a carbonyl of vanadium would have the formula $V_2(CO)_{12}$. Such a species has not been isolated, apparently due to steric hindrance. The unpaired electron in $V(CO)_6$ causes this molecule to be strongly paramagnetic.

The metal–carbon bond in metal carbonyls is usually quite labile, and the carbonyl groups can exchange readily with unbound carbon monoxide. Nickel tetracarbonyl, the first metal carbonyl to be reported (see Chapter 1), remains commercially important as part of the Mond Process for the purification of nickel, especially if cobalt is the contaminant. A stream of warm carbon monoxide, passed over the nickel, reacts to form the volatile

Table 8.1. Mononuclear Metal Carbonyls

Compound	Color	Melting point
$V(CO)_6$	Blue-green	Solid; d. 70°C
$Cr(CO)_6$	Colorless	154–55°C
$Mo(CO)_6$	Colorless	Solid; d. 150°C
$W(CO)_6$	Colorless	Solid; subl. 60–70°C/vac.
$Fe(CO)_5$	Yellow	-20°C; b.p. 103°C
$Ru(CO)_5$	Colorless	-22°C
$Os(CO)_5$	Colorless	-15°C
$Ni(CO)_4$	Colorless	-25°C

$Ni(CO)_4$. This is swept along to a region of higher temperature, where it decomposes. Cobalt does not react under these conditions and is left behind.

8.4 METAL CARBONYL IONS

Simple binary metal carbonyl ions are exclusively anions; carbonyl cations that have been reported have other ligands in addition to carbon monoxide.[9] Numerous anions of formula $M(CO)_n^{x-}$ (x = 1,2) have been reported[10,11] and are most often prepared by reduction of a neutral carbonyl with an alkali metal in an appropriate solvent, such as liquid ammonia:[11]

$$Co_2(CO)_8 + 2\ Na \xrightarrow{\ NH_{3(l)}\ } 2\ Na^+\ Co(CO)_4^- \qquad (8.1)$$

The stable metal carbonyl anions all follow the E.A.N. Rule, as can be seen in the isoelectronic series $V(CO)_6^-$, $Cr(CO)_5^{2-}$, $Mn(CO)_5^-$, $Fe(CO)_4^{2-}$, $Co(CO)_4^-$ and $Cu(CO)_3^-$. In these species, the transition metals have formally negative oxidation states. Since they have relatively low electronegativities, these metals are quite reactive when present as carbonyl anions, and undergo a variety of substitution reactions (L = monodentate ligand),

$$Na_2Cr(CO)_5 + L + 2\ H_2O \longrightarrow Cr(CO)_5L + 2\ OH^- + H_2 + 2Na^+ \quad (8.2)$$

$$NaMn(CO)_5 + ICN \longrightarrow cis\text{-}Na^+\ [Mn(CO)_4(CN)I^-] + CO \quad (8.3)$$

heterometallic insertions,

$$Na_2W(CO)_5 + Fe_2(CO)_9 \longrightarrow 2Na^+\ [(OC)_5W{-}Fe(CO)_4]^{2-} + Fe(CO)_5 \quad (8.4)$$

and coupling reactions.

$$NaCo(CO)_4 + (CH_3)_3SiCl \longrightarrow (CH_3)_3SiCo(CO)_4 + NaCl \quad (8.5)$$

This last reaction is widely used to form compounds containing metal–metal bonds.

8.5 METAL CARBONYL COMPOUNDS HAVING ONE-ELECTRON LIGANDS

8.5.1 Metal Carbonyl Hydrides

Ligands that form sigma bonds to transition metals are considered as "one-electron donors" under the E.A.N. Rule. Many halo derivatives are known, as are many such compounds containing hydrogen. These latter are extremely important as intermediates in industrial carbonylation reactions.

Neutral metal carbonyl hydrides are known (eg, $HCo(CO)_4$, $H_2Fe(CO)_4$), and some anionic species as well, eg, $HFe(CO)_4^-$. The hydrogen in these compounds usually dissociates as a proton:

$$HMn(CO)_5 \rightleftharpoons H^+ + Mn(CO)_5^- \qquad (8.6)$$

Cobalt tetracarbonyl hydride has been the most investigated of these compounds,[12] due to its importance for the oxo reaction, which may be represented by the equation

$$RCH=CH_2 + H_2 + CO \xrightarrow[\Delta]{Co_2(CO)_8} RCH_2CH_2\overset{\displaystyle O}{\overset{\displaystyle \|}{C}}H \qquad (8.7)$$

This equation is a simple summation of a process that involves several steps,[12] of which the initial ones are shown:

$$Co_2(CO)_8 + H_2 \rightleftharpoons 2\ HCo(CO)_4 \qquad (8.8)$$

$$HCo(CO)_4 \rightleftharpoons HCo(CO)_3 + CO \qquad (8.9)$$

$$HCo(CO)_3 + H_2C=CH_2 \rightleftharpoons$$
$$\left[\begin{matrix} 1:1 \\ \text{complex} \end{matrix}\right] \rightleftharpoons CH_3CH_2Co(CO)_3 \qquad (8.10)$$

The labile ethylcobalt intermediate then adds a carbon monoxide molecule to give a propionyl group, and eventually the final product, propionaldehyde, forms, with $Co_2(CO)_8$ being regenerated. The complex involves linkage of the ethylene pi-bond to the cobalt atom. For longer-chain olefins, isomerization becomes possible:

$$CH_3CH_2CH=CH_2 + HCo(CO)_3 \rightleftharpoons CH_3CH_2CH_2CH_2Co(CO)_3$$
$$+ CH_3CH_2\underset{\displaystyle CH_3}{\overset{\displaystyle |}{C}}HCo(CO)_3 \qquad (8.11)$$

$$CH_3CH_2\underset{\displaystyle CH_3}{\overset{\displaystyle |}{C}}HCo(CO)_3 \rightleftharpoons HCo(CO)_3 + CH_3CH=CHCH_3 \qquad (8.12)$$

The lability of the cobalt–carbon bond and the ready hydrocobaltation/dehydrocobaltation reactions in these systems gives rise to a wide variety of possible products. The oxo reaction[12,13] and the closely related Fischer-Tropsch reaction,[14] which involves reduction of metal-bound carbon monoxide to give organic products have an extensive variety of contemporary commercial applications.

8.5.2 Metal Carbonyl Alkyl Derivatives

These highly reactive compounds were the first sigma-bonded organo derivatives of transition metals (other than platinum or gold).[15,16] For the most part, these compounds have only one alkyl or aryl group per metal atom (eg, $CH_3Co(CO)_4$, $C_6H_5Mn(CO)_5$, etc.). The metal–carbon sigma bond is quite labile (Chapter 5.6). One very important class of reaction known for these species are the insertion reactions[17,18] (Chapter 2.3.3.5). For the reaction

$$RMn(CO)_5 + CO \longrightarrow R\underset{\underset{O}{\|}}{C}Mn(CO)_5 \tag{8.13}$$

investigations using labeled carbon monoxide indicated that the acyl carbonyl group formed from a carbon monoxide group already bonded to the manganese atom.[17] Carbon monoxide insertion is an important part of the oxo reaction and also in the titanium/aluminum catalysis of olefin polymerization (Chapter 5.5). Many molecules can undergo insertion reactions.[18] Most other reactions of these compounds are those typical of organometals containing strongly polar metal–carbon sigma bonds (Chapter 5).

8.5.3 *Metal Carbonyl Organometal Derivatives*

A variety of organometal(loid)s containing metal carbonyl groups are known. This category would include complexes between metal carbonyls and triorganophosphines (or -arsines, etc.), such as $R_3P—Cr(CO)_5$, or the anionic species $W(CO)_4(CH_2)_3P(C_6H_5)_2$, isolated as the tetraphenylphosphonium salt[19] and shown in Figure 8.2. Even the recently reported diphosphenes and diarsenes (Chapter 7.3.2) can form such complexes; examples of a diphosphene[20] and a diarsene[21] derivative of a metal carbonyl appear in Figure 8.2.

Many examples are also known of complexes containing direct metal–metal linkages, such as $(C_6H_5)_3SnCo(CO)_4$. These are generally prepared through coupling reactions, as shown in equation 8.5, and represent a class of compounds receiving increasing research attention.

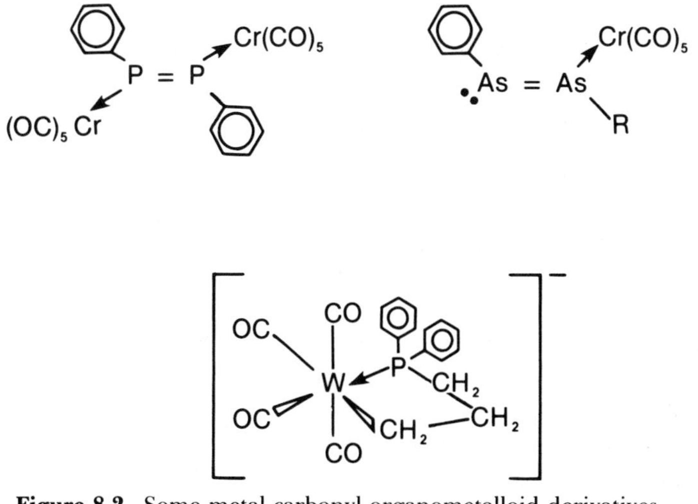

Figure 8.2. Some metal carbonyl organometalloid derivatives.

8.6 METAL CHALCOCARBONYLS

The chalcogen analogs of carbon monoxide—carbon monosulfide, carbon monoselenide and carbon monotelluride—have all been reported. These compounds are all highly reactive species, extremely difficult to isolate or retain under ordinary conditions. Carbon monosulfide has been detected in the upper portions of the terrestrial atmosphere and in stellar spectra. As with many unstable molecules, CS and CSe can form metal complexes considerably more stable than the molecules themselves.[22,23]

As of this writing, one binary metal chalcocarbonyl, $Ni(CS)_4$, has been reported.[24] This compound was prepared in an argon matrix at low temperatures; attempts to isolate it resulted in formation of a black solid, apparently a polymer.[24] Other chalcocarbonyl derivatives have one (occasionally two) CS or CSe group bonded to a metal that also has other ligands present. Some examples are $W(CO)_5CS$, $Cr(CO)_5CSe$ and $C_5H_5Mn(CO)_2CS$. Bridging CS groups are known, and will be discussed in Chapter 10 with bridging carbonyls. No complexes of CTe have yet been reported.

Generally speaking, carbon monosulfide in its complexes is both a better sigma donor and a better pi acceptor than carbon monoxide; consequently, the metal–carbon linkage in carbon monosulfide complexes is stronger, and less labile than the corresponding linkage in carbonyls. Preliminary evidence suggests that the same thing might be true in carbon monoselenide complexes.[22] Interestingly enough, infrared and Raman spectroscopic data indicate that the carbon–chalcogen stretching frequency in carbon monosulfide or monoselenide occurs at a higher value when the molecule bonds to a metal atom—exactly the opposite of what is observed for metal carbonyls! Doubtlessly more will be learned about these complexes in the years to come.

8.7 METAL ISOCYANIDES

As noted in Chapter 1, simple binary metal cyanides, $M(CN)_x$, are not considered as organometals, even though they may be isoelectronic with metal carbonyls. However, if an organic group is attached to the nitrogen, giving $M(CNR)_x$, the resulting compounds are considered to be organometals. This distinction appears to arise from custom rather than logic. The metal isocyanides (isonitriles) have an extensive chemistry,[25–27] at least in part because there are so many possible organic groups that might be used. In practice, methyl and phenyl isocyanide have been the most commonly studied. A few examples of the "parent" isocyanide, HNC, have been prepared,[28] such as $Cr(CO)_5CNH$. Isocyanides have a larger dipole moment than carbon monoxide, which contributes to their greater basicity. They frequently form complexes with metals in positive oxidation states, and there are isocyanide–metal compounds that have no carbonyl counterparts.

In its simplest form, the bonding in the carbon–nitrogen linkage may be described in a fashion analogous to that of the carbon–oxygen linkage in carbon monoxide. There are limits to the analogy, however, and the presence of the organic group frequently causes substantial differences,[26,27] including the existence of complexes where the M—C—N—R framework is bent.[27]

Much of the chemistry of metal isocyanide complexes corresponds to metal carbonyl chemistry, such as insertion reactions:

$$(CH_3)_3TaCl_2 \ + \ 3:CNC_4H_9\text{-t} \ \longrightarrow \ Cl_2Ta(\underset{\underset{CH_3}{|}}{-C}=NC_4H_9\text{-t})_3 \quad (8.14)$$

When complexed to a metal atom, the isonitriles become more reactive, since the carbon–nitrogen linkage can undergo addition reactions. This makes such complexes useful intermediates in organic syntheses. Treatment of metal isocyanides with alcohols can generate metal-bonded carbenes:

$$M—C{\equiv}N—R \ + \ R'OH \ \longrightarrow \ M{=}C\overset{\displaystyle\diagup NHR}{\diagdown OR'} \quad (8.15)$$

A similar method[27] may be used to prepare metal carbynes, such as $HFe_3(CO)_{10}C—N(CH_3)_2$.

8.8 METAL ALKYLIDENES

Numerous transition metal compounds containing carbenes, :CRR' groups, as ligands have been reported. Such compounds were first reported by Fischer[29] and have been frequently reviewed.[30–34] Such species might be considered as "carbene" complexes (analogous to isocyanide or carbonyl complexes), or they might be considered as "alkylidenes"—derivatives of alkyl groups with an alpha hydrogen removed. Both viewpoints have merit.

The metal–carbon linkage resembles those found in metal carbonyls and metal isocyanides. The carbon atom bonded to the metal is sp²-hybridized, with the electron pair in one of the hybrid orbitals. This interacts with an empty orbital on the metal to form the sigma component of the synergistic bond. The empty p-orbital on the carbon can then receive electron density from the metal d-orbitals. If there are unsaturated bonds or atoms with lone pairs of electrons attached to the carbene carbon, these may also interact with this orbital. In some respects, these systems resemble the ylids mentioned in Chapter 7.3.1. The analogy can even be extended to the method of preparation:[32]

$$(RCH_2)_3MCl_2 + 2\ LiCH_2R \longrightarrow (RCH_2)_3M{=}CHR + RCH_3 \quad (8.16)$$

where R is a t-butyl group and M is tantalum or niobium.

The chemistry of metal alkylidenes is substantial and varied. There appears to be some chemical differences between the "carbene" complexes of Fischer[29] and the alkylidenes of Schrock.[32,33] Trimethyl-phosphine liberated the carbene[29] from $(OC)_5CrC(CH_3)OCH_3$, but did not react[32] with $(C_5H_5)_2CH_3TaCH_2$. This latter compound, however, reacted with trimethylaluminum to form $(C_5H_5)_2CH_3\overset{+}{Ta}CH_2Al(CH_3)_3$—a reaction analogous to the ylids. Metal alkylidenes have received attention because of their importance in catalysis, particularly the metathesis of olefins[33,35,36]

$$2\ RCH{=}CH_2 \xrightarrow{\text{cat.}} RCH{=}CHR + H_2C{=}CH_2 \quad (8.17)$$

Closely related to alkylidenes are the carbyne complexes, which have (at least nominally) a metal–carbon triple bond. These were also prepared by Fischer.[29] The metal–carbon bond length shows a very distinct shortening in these compounds; for example, the chromium–carbon(carbyne) distance is 169 picometers, as compared to 222 picometers for a chromium–carbon (sp^2-hybridized) bond.[29] Alkynes can undergo metathesis in the same way as alkenes, and alkylidyne complexes can cause such a reaction:[32]

$$(RO)_3W{\equiv}CR + C_6H_5C{\equiv}CC_6H_5 \longrightarrow (RO)_3W{\equiv}CC_6H_5 \\ + RC{\equiv}CC_6H_5 \quad (8.18)$$

The oxidation state of the metal seems to be important, as the complex $(OC)_5BrW{\equiv}CC_6H_5$ did not undergo such a reaction.[32]

The carbene complexes might be considered analogous in some ways to corresponding amines, although the latter have no empty orbital available. Comparable divalent silicon and germanium species, $R_2Si:$ and $R_2Ge:$, are known as reaction intermediates, and would also be expected to form metal complexes under favorable conditions.

REFERENCES

1. Cotton, F. A.; Wilkinson, G., "Advanced Inorganic Chemistry," 4th Ed., J. Wiley & Sons, New York: 1980, pp. 1049–1079.
2. Huheey, J., "Inorganic Chemistry," 3rd Ed., Harper & Row: New York, 1983, pp. 591–609.
3. Douglas, B. E.; McDaniel, D. H.; Alexander, J. J., "Concepts and Models of Inorganic Chemistry," 2nd. Ed., J. Wiley & Sons: New York, 1983, pp. 405–418.
4. Hoffmann, R., *Science*, **1981,** *211,* 995.
5. Kettle, S. F. A.; Paul, I., *Adv. Organomet. Chem.*, **1972,** *10,* 199.
6. Sidgwick, N. V., "The Electronic Theory of Valency," Cornell University Press: Ithaca (NY), 1927, pp. 163–184.
7. Edgell, W. F.; Wilson, W. E.; Summitt, R., *Spectrochim. Acta*, **1963,** *19,* 863.
8. Jones, L. H.; McDowell, R. S., *Spectrochim. Acta*, **1964,** *20,* 248.

9. Abel, E. W.; Typfield, S. P., *Adv. Organomet. Chem.*, **1970**, *8*, 11.
10. Hieber, W., *Adv. Organomet. Chem.*, **1970**, *8*, 1.
11. Behrens, H., *Adv. Organomet. Chem.*, **1980**, *18*, 1.
12. Orchin, M., *Accts. Chem. Res.*, **1981**, *14*, 259.
13. Pruett, R. L., *Adv. Organomet. Chem.*, **1979**, *17*, 1.
14. Masters, C., *Adv. Organomet. Chem.*, **1979**, *17*, 61.
15. Heck, R. F., *Adv. Organomet. Chem.*, **1966**, *4*, 243.
16. Parshall, G. W.; Mrowca, J. J., *Adv. Organomet. Chem.*, **1968**, *7*, 5043.
17. Wojcicki, A., *Adv. Organomet. Chem.*, **1973**, *11*, 88.
18. Wojcicki, A., *Adv. Organomet. Chem.*, **1974**, *12*, 32.
19. Darensbourg, D. J.; Kudarski, R.; Delord, T., *Organometallics*, **1985**, *4*, 1094.
20. Borm, J.; Zsolnai, L.; Huttner, G., *Angew. Chem.*, **1983**, *95*, 1018.
21. Cowley, A. H.; Lasch, J. G.; Norman, N. C.; Pakulski, M., *Angew. Chem.*, **1983**, *95*, 1019.
22. Butler, I. S., *Accts. Chem. Res.*, **1977**, *10*, 359.
23. Rajan, S., *J. Sci. Ind. Res.*, **1979**, *38*, 648.
24. Yarborough, L. W.; Calder, G. V.; Verkade, J. G., *J. Chem. Soc., Chem. Comm.*, **1973**, 705.
25. Vogler, A., in "Isonitrile Chemistry" (Ugi, I., ed.), Academic Press: New York, 1971, pp. 217–234.
26. Treichel, P. M., *Adv. Organomet. Chem.*, **1973**, *11*, 21.
27. Singleton, E.; Oosthuizen, H. E., *Adv. Organomet. Chem.*, **1983**, *22*, 209.
28. Guttenberger, J. F., *Chem. Ber.*, **1968**, *101*, 403.
29. Fischer, E. O., *Adv. Organomet. Chem.*, **1976**, *14*, 1.
30. Gaspar, P. P.; Herold, B. J., in "Carbene Chemistry," (W. Kirmse, ed.), 2nd Ed., Academic Press: New York, 1971, pp. 504–550.
31. Schmidbaur, H., *Adv. Organomet. Chem.*, **1976**, *14*, 205.
32. Schrock, R. R., *Accts. Chem. Res.*, **1979**, *12*, 98.
33. Schrock, R. R., *Science*, **1983**, *219*, 13.
34. Casey, C. P., *React. Intermed.*, **1985**, *3*, 109.
35. Grubbs, R. H., in "Comprehensive Organometallic Chemistry," Pergamon: London, 1982, Vol. 8, pp. 500–551.
36. Dotz, K. H., et al, ed., "Transition Metal Carbene Complexes," VCH Publishers: Weinheim, 1983.

Chapter 9

Metal–Carbon Synergistic Bonds. II. Mononuclear Complexes of Unsaturated Hydrocarbons

9.1 INTRODUCTION

The first example of a compound containing an unsaturated hydrocarbon bonded to a metal was Zeise's Salt,[1] K^+ $C_2H_4PtCl_3^-$. Various other compounds of this type were prepared during the succeeding eleven decades, but the nature of the metal–hydrocarbon linkage was not fully understood until the preparation of ferrocene in 1951. This compound, $(C_5H_5)_2Fe$, was independently reported by two research groups[2,3] and has stimulated a flood of research activity that continues unabated after a third of a century.

For purposes of this discussion, we will subdivide the hydrocarbons into three groups: acyclic olefins and acetylenes; nonaromatic cyclic olefins; and aromatic hydrocarbons. The great majority of derivatives in the first two categories (there are some exceptions) have a single hydrocarbon bonded to the metal atom. Other ligands may be sigma-bonded inorganic species (anions or donor molecules), sigma-bonded organometal(loid)s, and synergistically bonded species (most commonly carbon monoxide). Metal–aromatic complexes usually have two per metal atom, although species with one aromatic hydrocarbon per metal are frequent. Very few compounds of any type having three or more unsaturated hydrocarbons bonded synergistically to a metal atom are known.

When named as ligands, unsaturated hydrocarbons are generally named in the usual fashion. Often, however, especially with polyolefins, there can arise an ambiguity as to how many of the double bonds are actually linked to a metal atom. To remove such an ambiguity, a special system has been devised.[4] The prefix "hapto" (from a Greek word meaning 'to fasten') indicated by the Greek letter eta, η, (sometimes a lower-case "h" has been used) and a numerical superscript indicating the number of carbon atoms bonded to the metal, is placed before the name of the hydrocarbon. Thus the compound $(\eta^6\text{-}C_6H_6)_2Cr$ would be named bis (η^6-benzene)-chromium(0), and the benzene is considered a hexahapto ligand. Hydrocarbons bonded to metals solely through synergistic metal–carbon bonds will almost always be named as even-numbered hapto species: dihapto, tetrahapto, hexahapto, etc. Hydrocarbons that combine a metal–carbon sigma bond with one or more metal–carbon synergistic bonds will be named as odd-numbered hapto species: trihapto, pentahapto, heptahapto, etc. Saturated alkyl groups or other hydrocarbons bonded to metals solely by sigma bonds are considered monohapto ligands. The symbol is included in the name or the formula only when it is necessary or desired to indicate the number of metal–carbon linkages present.

Trivial names are relatively infrequent. Zeise's salt and Zeise's dimer, named after their discoverer, owe their continuing use to historical custom. The compound $(\eta^5\text{-}C_5H_5)_2Fe$ has been known as ferrocene since its initial report, and its dominant importance has caused related species to be named by extension or analogy. Compounds of general formula $(C_5H_5)_2M$ are generically termed metallocenes and individually named according to the metal: cobaltocene, manganocene, vanadocene, etc. Derivatives having groups

$$\overset{\displaystyle O}{\overset{\displaystyle \|}{}}$$

on the rings are also named by this method: $CH_3\overset{O}{\overset{\|}{C}}C_5H_4(C_5H_5)Fe$, acetyl-ferrocene. Derivatives with inorganic groups bonded to the metal are named in similar fashion: $(C_5H_5)_2TiCl_2$, titanocene dichloride.

9.2 THE METAL–CARBON LINKAGE

The principles of the synergistic metal–carbon bond, introduced in the preceding chapter apply to olefin/acetylene complexes as well. The sigma component of the linkage (shown in Figure 9.1) involves interaction between an empty metal orbital (usually a dsp^2 or d^2sp^3 hybrid orbital) and the pi-electron cloud of the carbon–carbon multiple bond. The pi component forms by interaction of a metal orbital (of t_{2g} symmetry) having electron density with the pi* antibonding orbital of the multiple bond. When such a linkage forms, both components weaken the carbon–carbon bond. As with the carbonyls, this weakening can be detected and investigated through vibrational spectroscopy. The weakened bond becomes more susceptible to addition reactions; for this reason, transition metals are used in a wide variety of laboratory or industrial applications as catalysts, eg, hydrogenation:

$$C_2H_4 + H_2 \xrightarrow{\text{Pd}} C_2H_6 \qquad (9.1)$$

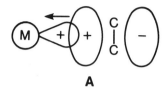

A

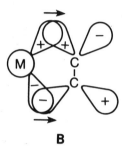

B

Figure 9.1. Orbital diagram of synergistic metal–carbon bond in metal–olefin compounds. (A) Donation from olefin to metal; (B) Donation from metal to olefin.

The same principles apply to polyenes bonded to metals, except that interactions between the double bonds must also be considered. The bonding of conjugated dienes to metals has been reviewed.[5,6] The bonding in ferrocene has been exhaustively investigated.[7,8] Since delocalized orbitals and more than two electrons are involved in such complexes, the bonding between aromatic and hydrocarbons is usually stronger than for simple olefins, and such linkages tend to be less labile.

Bonding in such complexes can usually be explained in terms of the E.A.N. Rule. Each pi-bond is considered as donating two electrons. Thus for ferrocene, if iron is considered to be present as Fe^{2+} (six electrons) and each cyclopentadienide anion donates six electrons, the total comes to eighteen. The corresponding cobalt compound, cobaltocene, would give nineteen electrons, but this compound readily oxidizes to give the cobalticinium ion, $(C_5H_5)_2Co^+$, which has eighteen. Mixed carbonyl–olefin complexes may be explained in similar fashion. In η^4-butadieneiron tricarbonyl, $C_4H_6Fe(CO)_3$, the iron has eight electrons, the butadiene provides four, and the three carbonyl groups provide six, making a total of eighteen.

9.3 STRUCTURES OF METAL–HYDROCARBON COMPLEXES

Development of structural techniques has paralleled the development of metal–unsaturated hydrocarbon complexes, and frequently contributed to it. Figure 9.2 shows certain representative structures. The anion

$C_2H_4PtCl_3^-$ has a square planar structure characteristic of Pt(II) complexes, with the ethylene molecule perpendicular to the plane of the $PtCl_3$ group and the midpoint of the C—C axis on the line formed by the platinum and the *trans*-chloride. The number and symmetry of the d-orbitals on platinum permit the free rotation of the ethylene group around the Pt-ethylene axis without affecting the pi-component of the synergistic bond.

Bis (η^3-allyl)nickel(II), also shown in Figure 9.2, is representative of many allyl and related complexes. The three carbons of each allyl group are coplanar, and the two planes are parallel. The nickel lies on the line between

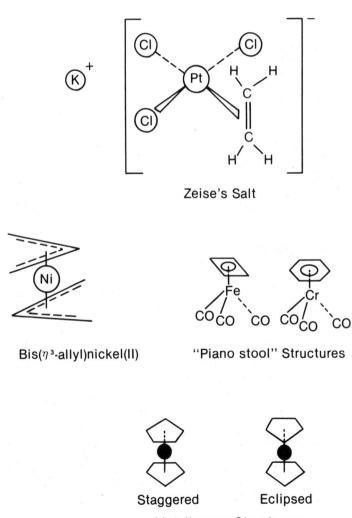

Zeise's Salt

Bis(η^3-allyl)nickel(II)   "Piano stool" Structures

Staggered   Eclipsed

Metallocene Structures

Figure 9.2. Some structures of mononuclear complexes of unsaturated hydrocarbons.

the geometric centers of the two triangles and is midway between them. Another common type of structure involves bonding between an olefin with at least two double bonds and a metal that also has three carbonyl groups bonded to it. In these "piano stool" structures, the plane of the carbonyl carbons lies parallel to the plane of the carbons actually linked to the metal atom.

Simple metallocenes usually have one of two structures shown in Figure 9.2. In both cases, the two rings are parallel to each other, and the metal atom is halfway between them (hence the commonly used term "sandwich structure"). The two structures differ by the orientation of the rings relative to each other. In the more common "eclipsed" form, the carbon atoms of the upper ring are directly above the carbon atoms of the lower ring; in the "staggered" form (as found in ferrocene), the carbon atoms of the upper ring lie directly above the midpoint of the carbon-carbon axes of the lower ring. If covalent groups are bonded directly to the metal atom, then the two rings no longer are parallel; in $(\eta^5\text{-}C_5H_5)_2TiCl_2$, the midpoints of the two rings lie at the vertices of a distorted tetrahedron. Bridging groups connecting the two rings of a metallocene may cause deviations from a completely parallel position. An unusual example of such a structure has been reported,[9] in which the two cyclopentadienide rings lie at the ends of a continuous helix of fused benzene rings.

Structures in the solid state are often constrained and sometimes altered by the fact that they occur in a rigid, even procrustean, lattice. Because of this, caution needs to be exercised when extrapolating from a solid-state structure to the same compound as a vapor or in solution. Frequently compounds will change their structure upon dissolution; in fact, many hydrocarbon–metal complexes will actually change from one structure to another after being completely dissolved. If such stereochemical alterations occur between structures that are chemically equivalent, the molecules undergoing such changes are termed fluxional molecules. Such behavior should be distinguished from tautomerism, in which the forms involved are not chemically equivalent.

Fluxional behavior is analogous to the rapid exchange of alkyls (Chapter 4.3) and may be studied by similar techniques. One of the earliest such systems was cyclooctatetraeneiron tricarbonyl, $C_8H_8Fe(CO)_3$. The structure suggested by proton NMR spectroscopy and that determined by x-ray crystallography seemed to be in conflict.[10,11] The former gave a single absorption peak, indicating that at room temperature all the hydrogens were equivalent. Analysis of the crystal structure, however, indicated a puckered hydrocarbon ring. Temperature-dependent proton NMR studies showed that, at room temperature, there is rapid interconversion among various structures in which only two double bonds at any one instant of time were actually bonded to iron.

Many, perhaps even a majority, of synergistic metal-hydrocarbon complexes show fluxional behavior. One such system, shown in Figure 9.3,

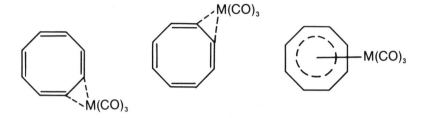

Cyclooctatetraene—Metal Tricarbonyl Complexes

Exo—Endo Interconversion

"Ring—Whizzing"

Figure 9.3. Examples of fluxional organometallic compounds.

involves exchange between the *exo* and the *endo* forms of an unsymmetric olefin; the relative contributions of the two forms will vary, depending on the nature of the olefin substituent(s), the metal, temperature, etc. Nor is fluxional behavior limited to synergistic metal–carbon linkages either. Many representative elements bonded to a cyclopentadiene ring undergo a ready migration around the ring. This process, sometimes termed "ring-whizzing," involves rapid breaking and reformation of the metal–carbon sigma bonds, rendering the electronic environment of all ring hydrogens equivalent.

9.4 CHEMICAL REACTIONS OF HYDROCARBON COMPLEXES

9.4.1 Substitution on the Hydrocarbon

Most substitution reactions have been reported for aromatic complexes. For the most part, such reactions show little difference from the corresponding reactions for uncomplexed hydrocarbons, although the presence of the metal usually alters the rate of reaction. Ferrocene has been most investigated in this respect, and some typical reactions are shown in Figure 9.4. Lithioferrocene is frequently used as starting material, simply because the lithium–carbon bond is so reactive that it will react readily with most reagents. Chloro- and nitroferrocene must be prepared from this compound rather than from ferrocene itself, as direct chlorination or nitration, involving Cl^+ or NO_2^+ as reaction intermediates, will cause oxidation of ferrocene to ferricinium ion.

Other aromatic complexes will also undergo substitution reactions such as Friedel-Crafts acylation:

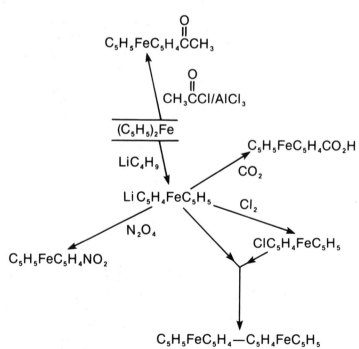

Figure 9.4. Preparation of some substituted ferrocenes.

$$C_4H_4Fe(CO)_3 + CH_3\overset{O}{\overset{\|}{C}}Cl \xrightarrow{AlCl_3} CH_3\overset{O}{\overset{\|}{C}}C_4H_3Fe(CO)_3 + HCl \quad (9.2)$$

$$(C_6H_6)_2Cr + CH_3\underset{\underset{O}{\|}}{C}Cl \xrightarrow{AlCl_3} CH_3\underset{\underset{O}{\|}}{C}C_6H_5CrC_6H_6 + HCl \quad (9.3)$$

Disubstituted products can also be prepared; these generally have one substituent on each ring. Substituents bridging between the rings are also known; one such compound of particular interest is (1,1'-ferrocene-diyl)dichlorosilane, $C_5H_5FeC_5H_4$—$SiCl_2$—$C_5H_4FeC_5H_5$, which has been used to coat the surfaces of photoelectric cells.[12] An even more unusual ferrocene with bridging groups is [1.1] ferrocenophane, in which the "upper" rings of two separate ferrocene molecules are joined by a methylene group, and the two "lower" rings are joined by a second bridging methylene group. This compound liberates hydrogen from acidic solutions, undergoes rapid hydrogen exchange on the methylene groups, and, as its carbanion, shows a carbon–hydrogen–carbon hydrogen bond.[13]

9.4.2 Exchange on the Metal Atom

Metal complexes of olefins or aromatics having other substituents on the metal atom can undergo exchange reactions. For the most part, these are the same as for inorganic derivatives of the same metals. As mentioned in the preceding chapter, carbonyl groups can undergo exchange. Hydrogen bonded to the metal can be used for addition reactions. Simple olefins bonded to metals may undergo displacement reactions:

$$C_2H_4PtCl_3^- + 4\ CN^- \longrightarrow Pt(CN)_4^{2-} + C_2H_4 + 3\ Cl^- \quad (9.4)$$

This reaction can be used to replace one olefin with another.

The metal atom in various metallocenes may undergo oxidation:

$$(C_5H_5)_2Co + Cl_2 + AlCl_3 \longrightarrow (C_5H_5)_2Co^+\ AlCl_4^- \quad (9.5)$$

The readiness with which this reaction will occur depends on the metal. Cobaltocene is more readily oxidized than ferrocene, because the resulting $(C_5H_5)_2Co^+$ ion has 18 electrons, whereas the ferricinium ion has 17. Oxidative addition to metals having hydrocarbon ligands can also occur. The majority of these involve alkyl iodides, but there have been a few examples of hydrocarbons undergoing oxidative addition:[14]

$$\eta^5\text{-}C_5H_5Re\ [P(CH_3)_3]_3$$

$$+\ C_6H_6 \xrightarrow{h\nu} [(CH_3)_3P]_2ReH(\eta^5\text{-}C_5H_5)C_6H_5 + (CH_3)_3P \quad (9.6)$$

This reaction replaces the rather sluggish carbon–hydrogen bond with the reactive rhenium–hydrogen linkage and the even more reactive rhenium–carbon bond, thereby in effect "activating" the carbon–hydrogen linkage. Such behavior is also known for rhodium and iridium complexes, and is of considerable interest to researchers. Cyclohexane did not undergo this reaction and made a very convenient solvent.[14]

9.4.3 Catalysis by Metal–Olefin Complexes

The importance of many metal–olefin complexes as intermediates in catalyzed syntheses has stimulated extensive research into their properties. Two examples have previously been mentioned. In the Ziegler-Natta catalysis of polymer formation (see Chapter 5.5), the binding of an olefin to the titanium weakens the carbon-carbon double bond, making it more susceptible to attack, and fixes it in space, which also facilitates reaction. In the Oxo Reaction (Chapter 8.5.1), the binding of the olefin to cobalt makes it vulnerable to hydrometallation. A third example is the Reppe Process:

$$RC{\equiv}CR + H_2O + CO \xrightarrow[HX]{Ni(CO)_4} cis\text{-}RHC{=}CR(CO_2H) \qquad (9.7)$$

The first step in a proposed mechanism for this process[15] is the displacement of one carbon monoxide molecule of nickel tetracarbonyl by the acetylene to form the complex $R_2C_2Ni(CO)_3$, which subsequently undergoes various reactions and rearrangements to form the product. The overall reaction is, in effect, the addition of a molecule of formic acid to one of the pi-bonds of the acetylene.

Another reaction involving nickel–acetylene intermediates in synthesis is the cycloaddition of four acetylene molecules to form cyclooctatetraene:[16]

$$4\ HC{\equiv}CH \xrightarrow{Ni} C_8H_8 \qquad (9.8)$$

While several mechanisms have been proposed, the most probable one involves an intermediate having four acetylenes bonded synergistically to nickel; these then undergo addition, either stepwise or concerted (a "zipper" mechanism) to form the product.

Numerous other examples of such catalyses are known; the interested reader is referred to Volume 17 of *Advances in Organometallic Chemistry*, which is exclusively devoted to this topic, or to books by Parshall[17] or Kochi.[18]

9.5 METAL COMPLEXES OF HETEROCYCLIC UNSATURATED MOLECULES

The examples heretofore cited for synergistic complexes of metals have been exclusively hydrocarbons. Numerous heterocyclic aromatic molecules are known, and certain of these will also form synergistic complexes with metals. These are less common and less well-known than hydrocarbon complexes. For molecules where the hetero atom is an element of Group

V or VI (eg, pyridine, thiophene), the tendency is for the hetero atom to form a simple donor-acceptor complex to the metal; numerous such complexes are known for pyridine.

Cyclic boron–nitrogen compounds form numerous complexes with metals.[19] These strongly resemble the corresponding hydrocarbons, and there are three-, four-, and five-electron donors known. One such complex is the molecule, $CH_3BC_5H_5CoC_4H_4$, which has a cyclobutadiene group also present.[20]

Heterocyclics with other elements have been less studied. The compound 2-methyl-l-phosphacyclopentadienylmanganese tricarbonyl is known,[21] and undoubtedly others will be reported as time goes on. The developing chemistry of unsaturated organometalloids should expand to include metal complexes within the near future.

REFERENCES

1. Thayer, J. S., *J. Chem. Ed.*, **1969**, *46*, 442.
2. Kealy, T. J.; Pauson, P. L., *Nature*, **1951**, *168*, 1039.
3. Miller, S. A.; Tebboth, J. A.; Tremaine, J. F., *J. Chem. Soc.*, **1952**, 632.
4. Cotton, F. A., *J. Am. Chem. Soc.*, **1968**, *90*, 6230.
5. Yasuda, H.; Tatsumi, K.; Nakamura, A., *Accts. Chem. Res.*, **1985**, *18*, 120.
6. Ernsy, R. D., *Accts. Chem. Res.*, **1985**, *18*, 56.
7. Lauther, J. W.; Hoffmann, R., *J. Amer. Chem. Soc.*, **1976**, *98*, 1729.
8. Hoffmann, R., *Science*, **1981**, *211*, 995.
9. Katz, T. J.; Pesti, J., *J. Am. Chem. Soc.*, **1982**, *104*, 346.
10. Jackman, L. M.; Cotton, F. A., ed., "Dynamic Nuclear Magnetic Resonance Spectroscopy," Academic Press: New York, 1975.
11. Faller, J. W., *Adv. Organomet. Chem.*, **1977**, *16*, 211.
12. Fox, J., *Chem. & Eng. News* (March 19, 1979), 25.
13. Mueller-Westerhoff, U. T.; Nazzal, A.; Proessdorf, W., *J. Am. Chem. Soc.*, **1981**, *103*, 7678.
14. Bergman, R. G.; Seidler, P. F.; Wenzel, T. T., *J. Am. Chem. Soc.*, **1985**, *107*, 4358.
15. Parshall, G. W., *Science*, **1980**, *208*, 1221.
16. Volhardt, K. P. C.; Colburn, R. E., *J. Am. Chem. Soc.*, **1981**, *103*, 6259.
17. Parshall, G. W., "Homogeneous Catalysis," Wiley: New York, 1980.
18. Kochi, J. K., "Organometallic Mechanisms and Catalysis," Academic Press: New York, 1978.
19. Siebert, W., *Adv. Organomet. Chem.*, **1980**, *18*, 301.
20. Herbevich, G. E.; Becker, H. J.; Hessner, B.; Zelenka, L., *J. Organometal. Chem.*, **1985**, *280*, 147.
21. Suryaprakash, N.; Kunwar, A. C.; Khetrapal, C. L., *J. Organometal. Chem.*, **1984**, *275*, 53.

Chapter 10

Metal–Carbon Synergistic Bonds. III: Polynuclear Compounds

10.1 INTRODUCTION

While the mononuclear complexes discussed in the last two chapters have been important in the development of synergistic organometallic chemistry, perhaps the majority of the most recent work in this area has involved polynuclear metal compounds. These contain two or more atoms which have carbonyl groups and/or unsaturated molecules bonded to them. The metals may be the same element or different elements; many examples of both types have been reported.

For the purposes of this discussion, polynuclear synergistic organometals will be subdivided into two categories: those with no metal–metal bonds; and those possessing metal–metal linkages. Compounds in the first category usually have the metal atoms held in position by bridging groups such as carbon monoxide, hydrocarbons, or (less commonly) sigma-bonded inorganic species. The metal–metal bonds in those polynuclear compounds possessing them may be sigma bonds (again, between atoms of the same metal or between atoms of different metals), or they may be multicenter bonds. In the latter situation, such compounds frequently occur in clusters of four or more atoms, with delocalized electrons binding the metal s; such compounds are currently receiving intense research investigation. Compounds with metal–metal bonds may also have bridging ligands.

10.2 COMPOUNDS HAVING NO METAL–METAL LINKAGES

Polynuclear synergistic organometals have no direct bond between the metal atoms always have some sort of bridging ligand. Carbon monoxide may serve this purpose, although bridging carbonyls are more usually found in those systems where metal–metal bonds are also present. Inorganic species may also serve as a bridge; hydrogen most commonly serves this purpose. The anion $Cr_2(CO)_{10}H^-$, shown in Figure 10.1(a), contains two chromium atoms which each have five terminal carbonyl groups and are linked together through hydrogen. A pair of bridging hydrogens occur in the compound $C_{20}H_{20}Ti$, whose structure is shown in Figure 10.1(b). Each titanium atom has a distorted tetrahedral configuration. In addition to the bridging hydrogens, there is a bridging fulvalene molecule as well.[1]

Numerous compounds containing bridging organic groups have been reported. Figure 10.1(c) shows a generalized chain structure containing

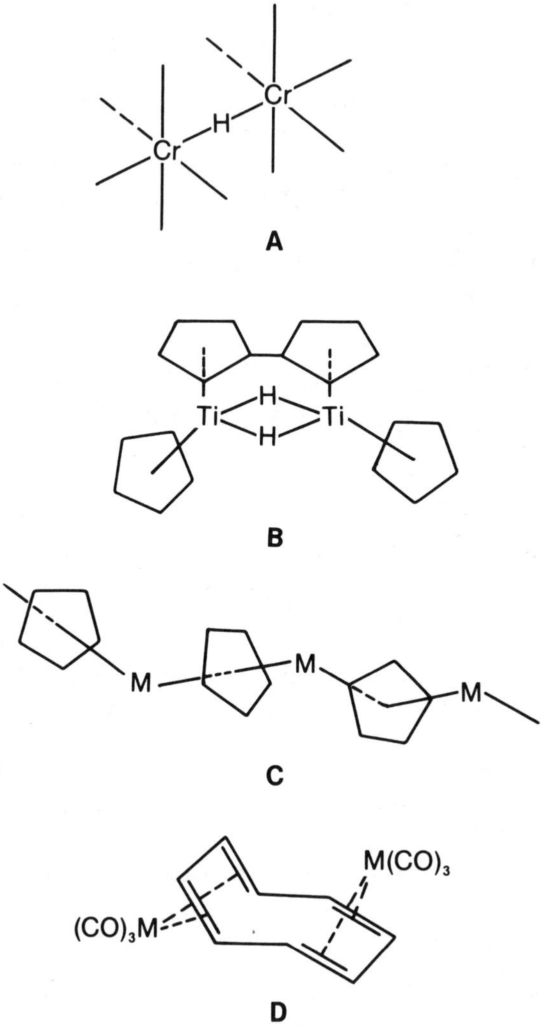

A

B

C

D

Figure 10.1. Structures of some polynuclear synergistic organometals having no metal–carbon linkages.

alternating cyclopentadienide groups and metals. The letter M may symbolize a lone metal atom, as in $[C_5H_5In]_x$, or a substituted metal, as in $(C_5H_5)_3Sc$, where M $= (C_5H_5)_2Sc$.[1] The rings may be joined, as in diferrocenyl, $C_5H_5FeC_5H_4$—$C_5H_4FeC_5H_5$, which has a bridging fulvalene group. Such linkages may continue indefinitely, as in polyferrocenyl. Metallocene derivatives containing bridging groups between rings are also known. One of these, [1.1] ferrocenophane, was mentioned in the preceding chapter, (9.4.1), and the hydrogens on the bridging methylene have unusual properties.

Nonaromatic polyolefins, both cyclic and acyclic, have the potential ability to bond to two different metal atoms, thereby becoming bridging ligands. One such example is shown in Figure 10.1(d), in which the four double bonds of cycoloctatetraene act as two separate η^4-units. Acetylenes can also act as bridging ligands, since the two pi-bonds are orthogonal to each other and can interact separately with different metal atoms. Some $\mu(\omega,\omega'\text{-al-kanediyl})$ metal complexes have been reported.[2] Certain of these, such as $(OC)_5Re$—CH_2CH_2—$Re(CO)_5$, have no metal–metal bonds.

A rather unusual type of bridging ligand is a germanium atom,[3] as found in the compound $R(OC)_2Mn=Ge=Mn(CO)_2R$, where R = pentamethylcyclopentadienyl. This species is proposed as having a manganese–germanium double bond.

10.3 POLYNUCLEAR COMPOUNDS WITH METAL–METAL BONDING

10.3.1 Compounds Lacking Bridging Groups

Synergistic polynuclear organometals with metal–metal bonds but having no bridging groups usually have single metal–metal sigma bonds. Dimanganese decacarbonyl, $Mn_2(CO)_{10}$, has the structure $(OC)_5Mn$—$Mn(CO)_5$, with each manganese atom showing octahedral symmetry. Triosmium dodecacarbonyl, $Os_3(CO)_{12}$, and its ruthenium counterpart, have the triangular structure shown in Figure 10.2(a). All carbonyl groups are terminal, and the metals are bonded by sigma bonds. Many compounds will have sigma bonds to representative metals, as in $(C_6H_5)_3SnCo(CO)_4$, and there are some species reported that have metal–metal multiple bonding.[4]

10.3.2 Oligonuclear Compounds Having Bridging Ligands

The majority of polynuclear compounds contain both metal–metal bonding and bridging ligands. This section will deal with "oligonuclear" compounds—species that have a relatively small number of metal atoms in linear, planar or cyclic structures. Cage polynuclear or cluster compounds will be discussed in the following section.

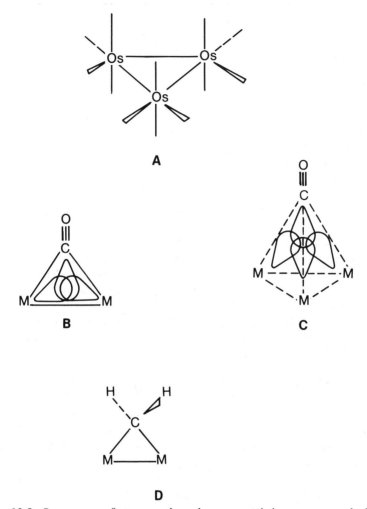

Figure 10.2. Structures of some polynuclear synergistic organometals having metal–metal linkages.

The first and the most studied of these bridging ligands is carbon monoxide. Figure 10.2(b) shows such a system in its most general form. The bonding follows the pattern described in Chapter 8, except that in these bridging systems, *both* metal atoms can donate electrons into the carbon–oxygen antibonding pi-orbital, causing a more extensive weakening of the bond and a larger shift to lower frequency for the carbon–oxygen stretching vibration. This is even more true for the "facial" carbonyls, shown in Figure 10.2(c), where a carbon monoxide molecule bonds to three metal atoms arranged in the triangular face of a polyhedron; facial carbonyls are found almost exclusively in cluster compounds. The carbon–oxygen stretching frequency occurs at 1900–2100 cm^{-1} for terminal carbonyls, 1700–1850 cm^{-1} for bridging carbonyls and 1600–1700 cm^{-1} for facial carbonyls.

Vibrational spectroscopy has long been used as a tool for determining the presence of terminal and/or bridging carbonyls in a particular molecule.[5] The molecule $Fe_3(CO)_{12}$ has a structure unlike its osmium or ruthenium counterparts; while the three iron atoms are arranged in a triangle, there are bridging carbonyl groups present as well as iron–iron bonds.

Another bridging ligand that has received considerable investigation is the methylene group.[2] This is shown in Figure 10.2(d). Theoretical investigations on such species apparently indicate that the M_2C system is best considered a "dimetallacyclopropane" analog. While there is synergistic interaction as described in Chapter 8, the bridging methylene group appears to be thermally more stable and chemically less labile than carbon monoxide.[2] The derived ligands vinylidene, $R_2C{=}C{:}$, and allylidene, $R_2C{=}C{=}C{:}$, are counterparts to isonitriles, and also show synergism in their bonding.[6]

Alkynes may likewise serve as bridging ligands, with the two pi-bonds interacting separately with two metal atoms that are also joined by a metal–metal bond. One example of this[7] is the molecule $R_2C_2Co_2(CO)_6$, where R_2C_2 is di-*tert*-butylacetylene. Inorganic groups may also serve as bridging ligands, although examples are less common than those discussed in Chapter 10.2. The anion $Fe_3(CO)_{11}H^-$ has two of the iron atoms joined together by a bridging carbonyl group, a bridging hydride ion and an iron–iron bond.

10.3.3 Organometallic Cage Compounds

Polynuclear compounds containing four or more atoms of one element bonded to each other usually form a closed polyhedral structure; such species are called "cage compounds." The previously mentioned carboranes (see Chapter 7.6) are one example. Numerous inorganic cage compounds of both main-group and transition elements have been reported; examples include neutral compounds B_4Cl_4, P_4O_6 and $Be_4O(CH_3CO_2)_6$, as well as the ions Bi_9^{5+} and $Mo_6Cl_8^{4+}$. However, the most-investigated examples of such compounds are the polynuclear transition of metal carbonyls.[8–12]

Metal atoms of cage compounds usually occupy the vertices of regular polyhedra.[13] These atoms interact with each other to form delocalized bonds (as do the boron atoms in carboranes), and the electron density of such bonds lies within the boundaries of the polyhedral structure, making such structures less susceptible to chemical fragmentation or thermal decomposition. Bridging carbonyl groups are almost always present; facial carbonyl groups may likewise be present. The anion $Ni_5(CO)_{12}^{2-}$, shown in Figure 10.3(a), has the five nickel atoms at the vertices of a trigonal bipyramid. There are nine terminal carbonyl groups (six at the two apical nickel atoms; one each at the three equatorial nickel atoms), and three bridging carbonyl groups in the equatorial positions. Similarly, the anion

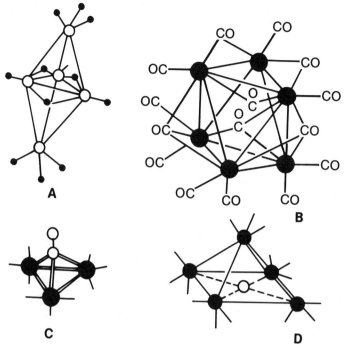

Figure 10.3. Some metal cluster compounds containing synergistic metal–carbon bonds.

$Co_6(CO)_{15}^{2-}$, shown in Figure 10.3(b), has the cobalt atoms in the form of a distorted octahedron, and terminal, bridging and facial carbonyl groups.

Various compounds having heterogeneous cage structures are also known. These might be all metals, as in the various mixed-metal cluster compounds of iron, cobalt, ruthenium and osmium,[9] or they might have a single non-metal or metalloid, usually in an apical position. Figure 10.3(c) shows a "facial" carbyne, as found in the compound $CH_3CCo_3(CO)_9$. The similar compound, $Fe_5C(CO)_{15}$, was the first example of a carbidocarbonyl.[14] In this compound the five iron atoms, each having three terminal carbonyl groups, lie at the vertices of a square pyramid, while the carbon atom lies some eight picometers below the basal plane, forming in effect a highly distorted octahedron. Other such carbidometal carbonyls have been reported.[15] The carbon atoms in these compounds are chemically reactive, and can form carbon–hydrogen or carbon–carbon bonds. The cage materials $RGeCo_3(CO)_9$ have been prepared,[16] and silicon analogs are also known.

In addition to their importance in catalysis (which will be considered in a subsequent section), cage compounds have an extensive chemistry.[7] They can react with various nucleophiles (particularly hydrogen) and with electron donors (particularly organic derivatives of phosphorus, sulfur, selenium, etc) to "cap" triangular faces by converting them to bases of irregular tetrahedra. Cage structures can alter their size and their composition. Conventional "ball and stick" structural models for cage synergistic organometals suggest a rigidity for such species that does not, in fact, exist; they frequently show marked fluxionality. Hexanuclear clusters might undergo interconversion among four possible geometric structures:[7] regular octahedron, trigonal prism, capped trigonal pipyramid, and capped tetragonal pyramid. In addition, a bicapped tetrahedron and capped square pyramid suggest themselves as possibilities. The series of compounds $[Pt(CO)_2]_{3n}^{2-}$ (n = 2 − 5) undergo ready interconversion among themselves, even at low temperatures.[7,17] The chemistry of organometal cage compounds offers a rich area for research, and is an extremely active part of current organometallic chemistry.

10.4 CATALYSIS AND POLYNUCLEAR ORGANOMETALS

The most important impetus for the investigation of polynuclear synergistic organometals, especially the cage compounds, involves their role in catalysis. This role has been reviewed in a series of articles.[18–26] Metal cage compounds form an important intermediate stage between a simple molecule possessing a single metal atom on the one hand, and a bulk metal catalyst having an extensive crystal array on the other. The largest cage species reported to date, $Pt_{19}(CO)_{22}^{4-}$, has a highly symmetric structure some 800 by 1100 pm in size, which approaches the dimensions of small platinum crystallites.[10] Certain physical properties, such as ionization potential or electron affinity, differ noticeably for an isolated metal atom in comparison to the same metal in bulk.[20] This observation suggests that metal clusters might owe their catalytic properties to electronic effects as well as structural effects.

One reaction of particular interest to catalytic chemists is the Fischer-Tropsch synthesis of hydrocarbons:[20,27]

$$n\,CO + 2n\,H_2 \longrightarrow (CH_2)_n + n\,H_2O \qquad (10.1)$$

Many catalysts have been studied for use in this synthesis, and the effects of substitution investigated. In the series of $Ir_4[(C_6H_5)_3P]_n(CO)_{12-n}$ where n may vary from one to three, the rate of hydrogenation of olefins decreased as the number of triphenylphosphine groups increased.[10] The flexibility and fluxional behavior of metal atoms in cluster or cage compounds, along with the mobility of ligands across the surface of cage structures will also play important roles in the potential applications as catalysts. Certain cage compounds have the ability to "activate" carbon dioxide, making them useful catalysts for the hydrogenation of a common, thermodynamically stable molecule:[28]

$$n\ CO_2 + 2n\ H_2 \longrightarrow (CH_2O)_n + n\ H_2O \qquad (10.2)$$

The exciting possibilities for the application of cage/cluster organometals as catalysts for reactions of industrial importance has greatly abetted research on these compounds.

REFERENCES

1. Maslowsky, E., *J. Chem. Educ.*, **1978**, *55*, 276.
2. Herrmann, W. A., *Adv. Organomet. Chem.*, **1982**, *20*, 159.
3. Korp, J. D.; Bernal, I.; Hoerlein, R.; Serrano, R.; Herrmann, W. A., *Chem. Ber.*, **1985**, *118*, 340.
4. Cotton, F. A., *Accts. Chem. Res.*, **1978**, *11*, 225.
5. Cotton, F. A.; Wilkinson, G., "Advanced Inorganic Chemistry," 4th ed., J. Wiley & Sons: New York, 1980, pp. 1070–1079.
6. Bruce, M. I.; Swincer, A. G., *Adv. Organomet. Chem.*, **1983**, *22*, 59.
7. Vahrenkamp, H., *Adv. Organomet. Chem.*, **1983**, *22*, 169.
8. Chini, P.; Longoni, G.; Albano, V. G., *Adv. Organomet. Chem.*, **1976**, *14*, 285.
9. Geoffrey, G. L., *Accts. Chem. Res.*, **1980**, *13*, 469.
10. Haggin, J., *Chemical & Engineering News* (Feb. 8, 1982), 13.
11. Cotton, F. A.; Chisholm, M. H., *Chemical & Engineering News* (June 28, 1982), 40.
12. Muetterties, E. L., *Chemical & Engineering News* (Aug. 30, 1982), 28.
13. Cotton, F. A.; Wilkinson G., *op. cit.*, pp. 56–60.
14. Braye, E. H.; Dahl, L. F.; Hubel, W.; Wampler, D. L., *J. Am. Chem. Soc.*, **1962**, *84*, 4633.
15. Bradley, J. S., *Adv. Organomet. Chem.*, **1983**, *22*, 1.
16. Gusbeth, P.; Vahrenkamp, H., *Chem. Ber.*, **1985**, *118*, 1770.
17. Brown, C.; Heaton, B. T.; Towl, A. D. C.; Longoni, G.; Fumagelli, A.; Chini, P., *J. Organometal. Chem.*, **1979**, *181*, 233.
18. Maugh, T. H., *Science*, **1983**, *219*, 474.
19. Maugh, T. H., *Science*, **1983**, *219*, 944.
20. Maugh, T. H., *Science*, **1983**, *219*, 1413.
21. Maugh, T. H., *Science*, **1983**, *220*, 592.
22. Maugh, T. H., *Science*, **1983**, *220*, 1032.
23. Maugh, T. H., *Science*, **1983**, *220*, 1261.
24. Maugh, T. H., *Science*, **1983**, *221*, 351.
25. Maugh, T. H., *Science*, **1983**, *221*, 1358.
26. Maugh, T. H., *Science*, **1983**, *222*, 151.
27. Haggin, J., *Chemical & Engineering News* (Oct. 26, 1981), 22.
28. Darensbourg, D. J.; Kudaroski, R. A., *Adv. Organomet. Chem.*, **1983**, *22*, 129.

Chapter 11

Organometallic Compounds in Biology. I: Medicinal and Biochemical Uses

11.1 INTRODUCTION

Virtually from their first appearance, organometallic compounds showed biological activity. Early workers (see Chapter 1.3) frequently had their health endangered and occasionally damaged from such species. Table 11.1 presents a chronological summary on this section of organometallic chemistry. The role of organometal(loid)s in the biological and environmental sciences has been presented in a recent monograph.[1]

The toxic effects of organometals led to their early use in medicine. Two compounds, mercurochrome, **1**, and merthiolate, **2**, originally developed during this period, remain in use to this day. Ehrlich's pioneering work on Salvarsan, **3** (also known as arsphenamine, "606," and various other names), opened up the field of chemotherapy. Salvarsan, and the related compounds Neosalvarsan, **4**, and Mapharsen®, **5**, remained the primary drugs for treatment of syphilis until the development of penicillin during the 1940's. Subsequent research on antibiotics emphasized purely organic compounds until recently, when there has been a resurgence of organometalloidal derivatives showing medicinal uses. Most of these remain in the investigative or testing stage.

1

2

TABLE 11.1. Chronological Summary of Research on Biological Interactions of Organometallic Compounds[a]

1760 —	Cadet prepares solution of methylarsenicals and notes toxic effects.
1837 —	Bunsen begins research on "Cadet's arsenical liquid." He isolates $(CH_3)_4As_2$, "cacodyl," and notes toxicity of this compound and its derivatives. About this time Gmelin and others begin reporting on "arsenic rooms."
1858 —	Buckton notes irritating effects of alkyltin compounds on mucous membranes.
1866 —	First reported fatality from poisoning by dimethylmercury.
1890 —	First isolation of nickel tetracarbonyl. This leads to development of the chemistry of metal carbonyls, with concomitant health hazards.
1891 —	Gosio reports that the volatile arsenic species found in "arsenic rooms" exists as an alkylarsenic compound.
1908 —	Ehrlich begins research on the antibiotic activity of aromatic arsenic compounds, with resulting isolation and application of Salvarsan. This research also lays the foundation for systematic chemotherapy.
1914 —	Beginning of World War I, which involves use of various organoarsenicals as poison gases. Subsequently, the compound Lewisite is prepared, which in turn leads to the development of B.A.L. and related antidotes.
1923 —	Development of tetraethyllead as a gasoline additive. Deaths of certain persons handling this compound lead to development of safe-handling techniques and physiological researches.
1933 —	Challenger reports formation of trimethylarsine by the action of molds on arsenious oxide. His subsequent research leads to the formulation of the concept of "biological methylation."
1954 —	"Stalinon" disaster in France. Around this time the first cases of Minamata Disease appear.
1961 —	The structure of Vitamin B_{12} coenzyme is shown to contain a cobalt–carbon bond, making it the first organometallic compound known to form as a standard product of biological metabolism.
1968 —	First report that methylmercuric compounds can be generated by the action of microorganisms on inorganic mercury compounds.

[a]Copyright 1984 by Academic Press Inc., and reproduced by permission of the copyright holder.

3

4

5

There has been a parallel even greater, growth, in the use of organo-metal(loid)s as tools for biochemical investigation. Some of this involves laboratory and clinical testing for potential medicinal or therapeutic uses, or investigations into the mechanism and parameters of such medicinal materials. Many laboratory applications, however, have no direct relationship to medicinal use. These use organometal(loid)s as reagents for specialized research involving cells, unicellular organisms, eukaryotes, or biochemical models of biological systems. Some of this work deals directly with the toxicology of organometal(loid)s and will be discussed in the next chapter.

11.2 ORGANOMETALLOID ANALOGS OF BIOLOGICALLY ACTIVE COMPOUNDS

Silicon and phosphorus lie right below carbon and nitrogen in their respective Periodic Groups. The importance of both carbon and nitrogen in biological chemistry has led many workers to investigate the effect of replacing a crucial atom of one of these elements by its heavier congenor. The analogy has been extended to include germanium and arsenic as well. Usually such a substitution causes a drastic change in the biological function and/or activity of the molecule. Choline, $(CH_3)_3\overset{\oplus}{N}CH_2CH_2OH$, a very important molecule, has had phosphorus, arsenic and antimony analogs prepared and studied. The arsenic analog, arsenocholine, occurs in many marine organisms (see Chapter 13), as does arsenobetaine, $(CH_3)_3\overset{\oplus}{As}CH_2CO_2{}^{\ominus}$. The boron analog, $(CH_3)_3\overset{\oplus}{N}\overset{\ominus}{B}H_2CO_2H$, in which a —$CH_2$— group is replaced by the isoelectronic —$\overset{\ominus}{B}H_2$— group, shows activity against cancer cells.

Silicon has been extensively investigated in this respect, and much has been learned about the activity of sila analogs of drugs.[2] Replacement of a quaternary carbon by silicon usually causes a change in the effect of the drug, but the change may be either a decrease or an increase in the activity. Trimethylsilyl esters of molecules having amino and/or hydroxyl groups have been extensively used for chromatographic separation and analysis procedures. Most work on sila drugs remains in the experimental stage.

The sulfonic acid group, —SO_3H, is frequently mentioned as an analog of the carboxylic acid group, —CO_3H, in that both might be considered as derivatives of sulfuric and carbonic acids respectively, with an organic group replacing a hydroxyl group. In the same way orthophosphoric acid would give the phosphonic acid group, —PO_3H_2. Numerous phosphonates show biological activity; phosphonoacetic acid, $H_3O_2PCH_2CO_2H$, an analog of malonic acid, has become an important antiviral drug. The various roles of phosphonates in biology have grown to the point where they have been reviewed in a monograph.[3]

The methylene group, —CH_2—, is isoelectronic with a divalent oxygen atom, —O—. Esters of pyrophosphoric(diphosphoric) and triphosphoric acids have the —O—PO_3H_2 linkage, and play a crucial role in various biological processes, including oxidative phosphorylation. Phosphonic acid analogs, having the —CH_2—PO_3H_2 group, have been prepared and investigated as substrates for kinetic and/or mechanistic studies. Methylene diphosphonate, $CH_2(PO_3H_2)_2$, can be utilized in the investigation of bone and tooth repair.

Finally, it should be mentioned that there are many organometalloids that are biologically active but which have no organic analogs. The organosilicon derivatives of triethanolamine, **6**, commonly termed "silatranes," show a wide range of effects, with the degree depending very much on the nature of the group *R*. Carbethoxygermanium sesquioxide, the anhydride of $HO_2CCH_2CH_2Ge(OH)_3$, which has no organic counterpart, shows extensive biological activity.

6

11.3 ORGANOMETAL(LOID)S AS ANTIBACTERIAL AND ANTIPARASITIC AGENTS

Salvarsan and Mapharsen, originally developed for the treatment of syphilis, have continued to be used to treat parasitic infections, such as canine heartworms—a growing problem of great concern to pet owners. Arsanilic acid, p-$H_2NC_6H_4AsO_3H_2$, and its derivatives now have widespread use as additives to poultry feed. Their function is to enhance the growth of the poultry, presumably by curbing bacteria that would otherwise cause harm.

Phosphonic acids have been isolated as metabolites of soil fungi (see Chapter 13), where they are believed to serve as inhibitors of competing bacteria. Some of these compounds have been used in medicine; the prime example of this is phosphonomycin(fosfomycin), **7**. Phosphonopeptides have also been prepared and studied for their antibacterial activity; the best known example of this is "alafosfalin"(L-alanyl-l-aminoethylphosphonic acid). These compounds are believed to owe their activity to interference with bacterial cell wall metabolism.

$$H_3C \diagdown \quad \diagup PO_3H_2$$

$$H \quad \diagdown \quad O \quad \diagup H$$

7

Mercurochrome and merthiolate, previously mentioned, remain in use as readily available skin antiseptic agents. Various preparations containing aryl derivatives of heavy metals in soaps, creams, liquids, mouthwashes, etc., can be used for the treatment of skin or mouth infections.[1] Such compounds, typified by phenylmercuric borate, are used exclusively for external application, and can be quite poisonous if taken internally. Aryl compounds are used because they are less toxic than alkylmetal counterparts. In France a preparation of ethyltin compounds (named "Stalinon") was prepared for internal consumption as a treatment for skin infections.[1] As a result one hundred persons died and an equal number were paralyzed.

11.4 ORGANOMETAL(LOID)S AS ANTIVIRAL AGENTS

Viral infections have long been a target for medicinal investigation and therapy. In recent years these infections have become more widely known, as witness the public concern over herpes and AIDS (acquired immune-deficiency syndrome), both caused by viruses. Phosphonoacetic acid, and its congenor, phosphonoformic acid, $H_2O_3P-CO_2H$, (an analog of oxalic acid), have been widely used in research as virucidal agents. These compounds act against the viral DNA polymerases, inhibiting their activity and thereby preventing reproduction. They are frequently used as standards for testing strains of virus. Unfortunately, resistant viral strains have appeared. Various other organometallic and organometalloidal compounds have been reported as showing antiviral activity. These compounds are still being tested in laboratories, but may eventually enter the pharmacopeia for treatment of virus infections.

11.5 ORGANOMETAL(LOID)S AS ANTITUMOR AGENTS

Numerous organometal(loid)s have shown antitumor activity in the laboratory, and some have moved on to clinical testing. Perhaps the most investigated of these materials is yet another phosphonic acid derivative: N-phosphonoacetyl-L-aspartic acid, **8**. A strong antitumor agent in its own right, this compound becomes even more potent in combination with 5-fluorouracil,[1] and is currently undergoing clinical testing. The isotope boron-10 has a very high neutron-capture cross-section. For this reason many organoboron compounds have been prepared and tested in an effort to take advantage of this property for cancer therapy.[4] 1,2-Dicarbacloso-dodecaborane-12 has been used for this purpose.[5]

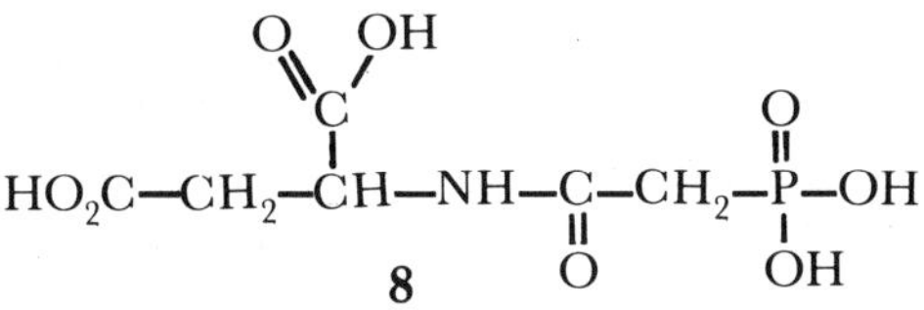

$$HO_2C-CH_2-\overset{\displaystyle \overset{O}{\underset{\|}{C}}\diagup^{OH}}{CH}-NH-\overset{\|}{\underset{O}{C}}-CH_2-\overset{O}{\underset{OH}{\overset{\|}{P}}}-OH$$

8

The great success of *cis*-$PtCl_2(NH_3)_2$, ("*cis*-diplatin") in cancer chemotherapy has stimulated investigation into other organometals. Metallocene dichlorides, $(C_5H_5)_2MCl_2$ (M = Ti, V, Nb, W, Mo), have shown strong activity against Ehrlich ascites tumors in mice,[1] and some ferricinium salts (but not ferrocene) also show antitumor activity.[6] The cyclic organogermanium compound, **9**, ("spirogermanium") has received extensive clinical testing as an antitumor agent. Carbethoxygermanium sesquioxide shows antitumor activity, as does its sulfur analog,[7] $(RGe)_2S_3$. In a structure-activity investigation on nitrogen complexes of dialkyltin dihalides, the antitumor action was found to depend on the tin-nitrogen bond length.[8] A conjugate between arsanilic acid and an immunotoxin showed activity against human pancreatic cancer.[9] A survey of organometallic compounds for carcinogenic activity recommended the following as models:[10] $Cr(CO)_6$, $(C_5H_5)_2Fe$, $(C_6H_5)_3Sb$, $(CH_3)_8Si_3O_2$, and $[(CH_3)_2SiO]_4$.

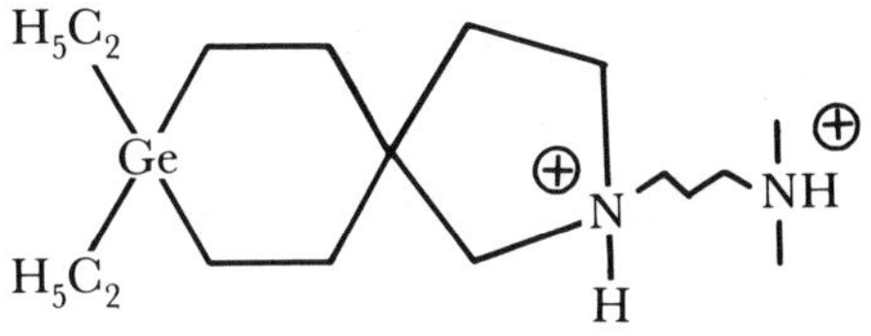

9

In view of the great public concern over cancer and the strong desire to find active anticancer agents, it seems likely that some of the many organometal(loid)s that show antitumor activity will eventually find their way into the pharmacopoeia.

11.6 ORGANOMETAL(LOID)S AS THERAPEUTIC AGENTS

11.6.1 Vitamin B₁₂

Vitamin B_{12}, originally known as the "antipernicious anemia factor," exists in several forms and has an extensive chemistry.[11,12] The form that exists in the body is known as the "coenzyme form" or by the common name "5'-

deoxyadenosylcobalamin." As shown by its structure, **10**, this compound has a direct cobalt–carbon sigma bond. Most of the chemistry of this enzyme involves the breaking of that bond to form a labile coordination site on cobalt, which, in turn, catalyzes a variety of reactions that involve the formation and breaking of transient cobalt–carbon bonds. Coenzyme B_{12} shows the right balance between stability and lability to make it suitable for biological purposes. A second important form is methylcobalamin, which has a methyl group in place of the 5′-deoxyadenosyl moiety in coenzyme B_{12}. Methylcobalamin is receiving attention in its own right as a pharmaceutical agent in a variety of situations,[1] including treatment of facial palsy[13] and tumors as an immunostimulating agent.[14]

10

The primary purpose of methylcobalamin, however, appears to be as a methylating agent; this aspect of its chemistry will be considered in Chapters 12 and 13. The medicinal form of Vitamin B_{12} is cyanocobalamin, which has a cyanide group attached to cobalt. In part this may be due to historical reasons, as cyanocobalamin was the first B_{12} derivative to be obtained in pure form, and can be readily recrystallized.

11.6.2 Other Therapeutic Organometal(loid)s

Various organometal(loid)s have been reported to have applications as therapeutic agents. A triethylphosphinegold(I) peracetylglucose compound (commonly known as "auranofin"), **11**, is presently being investigated in the treatment of rheumatoid arthritis. The siloxane $(CH_3)_3SiOSi(CH_3)_2CH_2SH$ in the form of an ointment may be used in the treatment of various skin disorders, such as scabies, eczema and psoriasis.[15] Various organophosphorus or organosilicon compounds have shown antiinflammation activity. Certain organomercurials have been used for many years as diuretic agents. Selected organometalloids can be applied to the treatment of both hypotension and hypertension. While there is considerable activity in this area,[1] most of it remains in the experimental stage. However, there seems every likelihood that many of these compounds will find use in therapy in the near future.

11

11.7 USES OF ORGANOMETAL(LOID)S IN BIOCHEMICAL INVESTIGATIONS

11.7.1 Interactions with Enzymes

The binding of organometals to active sites of enzymes is an important reason for their toxicity (see Chapter 12). Such compounds may also serve as tools for the investigation of such sites. Organoboron compounds act as reagents for binding to the proteolytic enzyme chymotrypsin (EC 3.4.21.1), and subsequent study of the resulting conformational changes.[1] Organo derivatives of heavy metals can deactivate enzymes by binding to their active sites. The reverse is also occasionally true; certain organomercurials are known to activate enzymes.[1] Triethyllead acetate binds to hemoglobin,[16] and affects certain brain enzymes.[17] Organophosphorus compounds having the P—CH_2—P linkage have been tested as substrates for enzymes that hydrolyze the P—O—P linkage.[18,19]

11.7.2 Organometal(loid)s as Biological Probes

Organometal(loid)s containing a radioactive metal or metalloid have served as very useful probes for biomedical investigations. Often these probes supplement studies using inorganic forms of the metal, as the solubility of metals in both water and lipids (hydrocarbons) depend very much on what ligands happen to be attached. The enormous concern arising from Minamata Disease (see Chapter 1.3) stimulated considerable research into the biological activity of methylmercuric compounds; this, in turn, has resulted in the use of $CH_3{}^{203}HgCl$ as an investigative tool. Similar uses have been found for derivatives containing ^{32}P, ^{74}As and ^{59}Fe. Various organotellurium(II) compounds containing ^{123m}Te have been used as myocardial imaging agents; for example, 15-(p-iodophenyl)-6-tellurapentadecanoic acid served for measurement of both occlusion and reperfusion flows in coronary investigations.[20]

The radioactive atom of an organometal(loid) need not be the metal or metalloid itself. The organic group may contain ^{14}C and/or ^3H (tritium). Investigations using $(C_6H_5)_3PCH_3{}^+$ salts with the methyl group having either or both of these labels have been reasonably frequent.[1] Finally, there are organometalloidal compounds in which a radioactive metal atom has no direct metal–carbon bonds but is coordinated to an organometalloid. The most common example is the use of methylenediphosphonates, $CH_2(PO_3H)_2{}^{2-}$, as carriers for ^{99m}Tc derivatives in bone-imaging experiments.

11.7.3 Organometal(loid)s and Membrane Investigations

In recent years, organometal(loid)s have served as useful tools for the measurement of cell membrane potentials and the transport of ions across cell walls. Organometalloidal ions, such as $(C_6H_5)_3PCH_3{}^+$, $(C_6H_5)_4P^+$, $(C_6H_5)_4As^+$ and $(C_6H_5)_4B^-$, have the uncommon combination of ionic charge, compact structures and lipophilic groups. Such ions dissolve in both water and hydrocarbons, making them especially useful for the investigation of membranes, which have both hydrophilic and hydrophobic (lipophilic) regions. Organoboronic acids, $RB(OH)_2$, can facilitate the transfer of water-insoluble materials across membranes. This process has become sufficiently well-known that the name "boradeption" has been proposed for it.[21]

Trialkyltin compounds have been used to investigate the transport of ions across cell membranes, with particular reference to the exchange of chloride and hydroxide ions.[22] They generally retard such exchange by inhibiting the membrane-bound enzymes that facilitate it. Triethyllead chloride behaves in a similar manner,[23] with the additional factor that it has considerable ability to permeate cell membranes, making it quite toxic.[24]

11.7.4 Organometal(loid)s and Cytological Investigations

Cytology is the study of cells, specifically cells within multicellular organisms: independently living cells (prokaryotes) will be considered in Chapter 12. Some work in this area has already been mentioned, most notably cancer research and the study of cell membranes. Numerous reports have appeared on cytotoxicological studies, with methylmercuric chloride being most commonly employed. Cytotoxic effects vary considerably, depending on such factors as the nature of the cell being used, the concentration of organometal(loid), the presence or absence of complexing agents, etc. Occasionally comparative studies are reported. In a study on mouse L-1210 leukemic cells, triphenyltin, and -lead chlorides, along with triphenylphosphinegold(I) chloride, showed marked cytotoxic effects.[25] Certain organometals, most notably methylcobalamin, stimulate cell metabolism and growth, making them potentially useful for therapy. l-Chlorosilatrane applied to wounded rat granular-fibrous tissue promoted growth and healing.[26] Organomercurials can be used for the investigation of cell division.[27]

11.7.5 Immunological and Genetic Studies Involving Organometal(loid)s

Landsteiner, in his pioneering work on antigens and antibodies,[28] used substituted phenylarsonic acid derivatives as reagents. Such compounds continue to be used for immunological and genetic investigations, especially p-azobenzenearsonic acid, $p\text{-}C_6H_5N{=}NC_6H_4AsO_3H_2$. Organomercurials and organotin compounds frequently caused disruption in organisms' immune responses; this has been reviewed for dialkyltin dihalides.[29] By contrast the previously mentioned (Chapter 11.2) compound carbethoxygermanium sesquioxide partially restores impaired immune response and is currently being investigated as an immune adjuvant. Part of this compound's anticancer activity probably arises from the stimulation it provides to the body's immune response.[30]

Organometal(loid)s have also been used for investigations into the functions of mitochondria; alkyltin compounds have been reported most frequently in this type of study.[1] In general, these compounds affect cell mitochondria in the same way they affect cell membranes, causing disruption of ion transport mechanisms and respiratory chains. Cyclopentadienylmanganese tricarbonyl, methylmercuric chloride, tetraphenylborate salts and phenylarsenic oxide also have been used in research involving mitochondria.

Organometal(loid) poisoning may cause mutagenic effects. Organomercurials are most commonly used here, although organoantimonials and organoleads have also been used. Previously mentioned phosphonic acids N-phosphonoacetyl-N-aspartic acid, 8, and phosphonoacetic acid have been used to determine whether viral resistance to these compounds is heredity; results indicate that such resistance may be only partially heritable.[1]

11.7.6 Miscellaneous Uses

Silicones have been used in various biochemical research projects for the properties that make them so important commercially—resistance to water and low surface adherence. Substitution of trimethylsilyl groups onto the oxygen or nitrogen atoms of biologically important molecules not only makes them more easily separated by chromatography, but also alters their biological activity. There is tremendous potential for the use of organometals and organometalloids in biochemical investigations, and this potential has only begun to be exploited.

REFERENCES

1. Thayer, J. S., "Organometallic Compounds and Living Organisms," Academic Press: New York, 1984.
2. Fessenden, R. J.; Fessenden, J. S., Adv. Organomet. Chem., 1980, 18, 275.
3. Hildebrand, R. L., ed., "The Role of Phosphonates in Living Systems," CRC Press: Boca Raton, Florida, 1983.
4. Barth, R. F.; Johnson, C. W.; Zei, W. Z.; Carey, W. E.; Soloway, A. H.; McGuire, J., Cancer Detect. Prev., 1982, 5, 315.
5. Goldenberg, D. M.; Sharkey, R. M.; Primus, F. J.; Mizusawa, E.; Hawthorne, M. F., Proc. Natl. Acad. Sci. USA, 1984, 81, 560.
6. Koepf-Maier, P.; Koepf, H.; Neuse, E. W., Angew. Chem., 1984, 96, 446.
7. Kakimoto, N.; Tanaka, N.; Miyao, K.; Ohnishi, T., Chem. Abstr., 1985, 103, 37622a.
8. Crowe, A. J.; Smith, P. J.; Hardin, C. J.; Parge, H. E.; Smith, F. E., Cancer Lett., 1984, 24, 45.
9. Runge, R. G., Chem. Abstr., 1985, 102, 67392x.
10. Doeltz, M. K.; Mackie, M.; Rich, P. A.; Lent, D.; Sigman, C. C.; Helmes, C. T., J. Environ. Sci. Health, 1984, 19A, 27.
11. Dolphin, D. H., ed., "B₁₂," 2 vol., Wiley: New York, 1982.
12. Pratt, J. M., "Inorganic Chemistry of Vitamin B₁₂," Academic Press: London, 1972.
13. Fujita, H.; Murakami, S.; Matsumoto, Y.; Kozawa, T.; Yanagihara, N., Biol. Abstr., 1985, 79, 43455.
14. Shimizu, N.; Hamazoe, R.; Kanayama, H.; Maeta, M.; Koga, S., Chem. Abstr., 1985, 103, 21568b.
15. Dvorak, M.; Resl, V.; Cermak, J.; Miksa, J., Chem. Abstr., 1984, 101, 28271s.
16. Aldridge, W. N.; Street, B. S., Toxicol. Appl. Pharmacol., 1984, 73, 350.
17. Bondy, S. C.; Hall, D. L., Chem. Abstr., 1985, 102, 91019b.
18. Engel, R., Chemical Reviews, 1977, 77, 349.
19. Holy, A., in "Phosphorus Chemistry Directed Towards Biologu," (Stec, W. J., ed.), Pergamon Press: Oxford, 1980, pp. 53–64.

20. Bianco, J. A.; Pape, L. A.; Alpert, J. S.; Zheng, M.; Hnatowich, D.; Goodman M. M.; Knapp, F. F., *Biol. Abstr.*, **1984**, *78*, 77904.
21. Gallop, P. M.; Paz, M. A.; Henson, E. B., *Science,* **1982,** *217*, 166.
22. Selwyn, M. J., in "Organotin Compounds: New Chemistry and Applications" (Zuckerman, J. J., ed.), Advances in Chemistry Series #157, American Chemical Society: Washington, 1976, pp. 204–226.
23. Stournaras, C.; Weber, G.; Zimmermann, H. P.; Doenges, K. H.; Faulstich, H., *Chem. Abstr.*, **1985,** *102*, 41239c.
24. Haeffner, E. W.; Zimmermann, H. P.; Hoffmann, C. J. K., *Chem. Abstr.*, **1985,** *102*, 41196m.
25. Yamamoto, Y.; Numasaki, Y.; Murakami, M., *Chem. Abstr.*, **1985,** *103*, 47866x.
26. Mansurova, L. A.; Voronkov, M. G.; Slutskii, L. I.; Dombrovska, L. E.; Bumagina, T. P., *Biol. Abstr.*, **1984,** *78*, 21877.
27. Thrasher, J. D., in "Drugs and the Cell Cycle," (Zimmerman, A. M.; Padilla, G. M.; Cameron, I. L., eds.), Academic Press: New York, 1973, pp. 25–48.
28. Landsteiner, K., "The Specificity of Serological Reactions," Dover: New York, 1962.
29. Penninks, A. H.; Seinen, W., in "Proceedings of the First International Symposium on Immunotoxicology," (Gibson, G. G.; Hubbard, R.; Parke, D. V., eds.), Academic Press: London, 1983.
30. Sato, I.; Yuan, B.; Nishimura, T.; Tanaka, N., *Chem. Abstr.*, **1985,** *103*, 64443z.

Chapter 12

Organometallic Compounds in Biology. II: Toxicological and Biocidal Aspects

12.1 INTRODUCTION

Organometallic compounds have been known to be toxic virtually since their first report (see Chapter 1.3). For the most part fatalities have been few and far between; however, there have been three widespread occurrences of what might be termed "epidemic poisonings by organometals": "Gosio-gas"(trimethylarsine), the toxic vapor of "arsenic rooms" that poisoned many persons in the 19th and early 20th Centuries (probably earlier as well); "Stalinon"(triethyltin iodide), which killed more than one hundred people in France around 1950; "Minamata disease" (methylmercuric compounds), which has afflicted hundreds of persons in Japan, Iraq, Guatemala and elsewhere. First reported around 1950, Minamata disease continues to occur. Industrial processes involving organometals, particularly the use of tetraethyllead as a gasoline additive, resulted in some workers being poisoned and led to the development of safety measures. Certain organoarsenicals were developed and occasionally used as poison gases in World War I; organophosphorus compounds have subsequently been prepared for the same purpose.

From these poisonings has grown a massive research interest in the extent, the effects and the mechanism of organometal(loid) toxicity towards humans, especially the alkyl derivatives of mercury, tin and lead. Such research has expanded to include other mammals and even other vertebrates, and has also sought the development of antidotes. Often these organometals have been used in biochemical investigations into neurons and related nervous tissue. The toxicity of these compounds has also led to their commercial development in a variety of biocidal applications against microorganisms, insects, aquatic invertebrates or plants. Organotin compounds have seen the most extensive use in this category.[1] Finally, organometalloids have become commercially important in agriculture for the control of plant growth and ripening.

12.2 ORGANOMETAL(LOID) TOXICITY TOWARDS MAMMALS

The very extensive current research into the toxic effects of organometal(loid)s towards humans and other mammals grew out of the poisonings described in the preceding section and also the need to develop safety precautions against poisonings caused by handling the commercially important organometals, such as tetraethyllead, triethylaluminum, cyclopentadienylmanganese tricarbonyl, nickel tetracarbonyl, etc. Such research has two closely linked primary goals: understanding the mechanism(s) by which such compounds are toxic, and developing appropriate methods of treatment, particularly antidotes. The first such antidote to receive widespread use was British Anti-Lewisite (B.A.L.; also known as "mercaprol"), $HOCH_2CH(SH)CH_2SH$; originally intended as an antidote for poisoning by Lewisite, $ClCH\!=\!CHAsCl_2$, B.A.L. has subsequently been used for treatment of poisonings by a variety of heavy metal compounds.[2]

Given the great diversity of organometallic compounds known and the even greater diversity of biochemical and metabolic processes present in all living organisms, one should not expect that any single mechanism would account for organometal toxicity.[3] As the reported research on this subject develops and becomes more sophisticated, the number of possible mechanisms increases. The most frequently proposed mechanism however, involves an organometal(loid) binding to the active site of an enzyme, thereby blocking its activity and disrupting a biochemical reaction sequence. Organophosphorus compounds, for example, combine with one of the active sites on the enzyme acetylcholinesterase. This enzyme controls the rate of the hydrolysis of acetylcholine

$$CH_3\overset{\displaystyle O}{\overset{\displaystyle \|}{C}}OCH_2CH_2\overset{+}{N}(CH_3)_3 + H_2O \longrightarrow HOCH_2CH_2\overset{+}{N}(CH_3)_3 + CH_3CO_2H \quad (12.1)$$

which in turn controls the rate of electrical impulses passing through the nervous system. Because of this disruption, compounds such as Sarin, $CH_3P(\!:O)(OC_3H_7)F$, are extremely potent neurotoxins. Organoarsenic(III), organomercury and other organo compounds of heavy metals bind strongly to thiol groups that comprise active sites of various enzymes such as dihydrolipoamide acetyltransferase. Trialkyltin compounds cause the inactivation of adenosyltriphosphatase bound to the surface of membranes, thereby causing the uncoupling of the process of oxidative phosphorylation.

Organometal(loid)s may bind to important sites on biologically important molecules that are not enzymes. An important example is myelin, a white fatty component of the central nervous system. Various organometals, particularly organotin compounds, cause this material to degenerate, thereby creating neural disorders. Organomercurials are known to cause systemic disorders, including teratological effects in fetuses.

The affinity of mercury and other heavy metals for sulfur suggested the use of sulfur compounds, especially thiols, as antidotes. Metallothionein, a protein with molecular weight of approximately 10,000, 30% of which consists of thiol-containing cysteine molecules, apparently serves this purpose in the body.[4] Mercaprol and D-penicillamine combine with heavy metals to form chelates which are stable and readily excreted. These compounds have proved less satisfactory for the treatment of poisoning by methylmercuric compounds. Selenium and its compounds, particularly sodium selenite, appear very promising in this area. One proposed mechanism has methylmercuric ion and selenite combining to form bis(methylmercuric) selenide, which rearranges in the following manner:

$$(CH_3Hg)_2Se \longrightarrow (CH_3)_2Hg + HgSe \qquad (12.2)$$

Mercuric selenide is water-insoluble and unreactive. Dimethylmercury is a volatile material, readily soluble in lipids and considerably less reactive than a monomethylmercuric species. Similar redistributions are reported for trimethyltin and trimethyllead selenides. Laboratory tests seem to indicate that many of the toxic effects of organometal poisoning are reversible if treated in time.

Not all toxic effects of organometal(loid)s are necessarily obvious or macroscopic. Treatment of small mammals with very low concentrations of organomercurials or organotin compounds cause behavioral disorders, such as memory impairment or inability to learn.[5] Such effects presumably arise from the effect of such compounds, even at low levels, on the nervous systems. Organometal(loid)s can also disrupt various body cycles, again even at low levels and without fatal effect. Mammals and other organisms can often detect the presence of organometals, and such compounds have potential use as repellants or antifeedants; for example, burlap bags treated with organolead compounds repelled rodents from the seeds within. Such applications have been more extensively investigated for insects, and will be discussed in Section 4 of this chapter.

12.3 ORGANOMETAL(LOID) TOXICITY TOWARDS NONMAMMALIAN VERTEBRATES

Vertebrates other than mammals have been less extensively investigated on their susceptibility towards organometal poisoning, and a greater proportion of the work that has been done has been carried out in the field rather than in the laboratory. The ability of various species of fish to accumulate and tolerate relatively high concentrations of methylmercuric compounds has been the focus of much research interest for many reasons, chief of which has been the use of fishes as human food. Fish that live in mercury-contaminated waters develop high levels of methylmercuric compounds in their tissues; this is called "bioaccumulation" (see Chapter 13). Such accumulation makes them poisonous to fish-eating predators, such as seals, cats, ospreys—or humans. The early cases of "Minamata disease" in Minamata Bay, Japan, developed in just this way.

Various species of birds can also develop poisoning by ingesting methylmercuric-containing food. Fish-eating birds, such as ospreys, frequently are poisoned in this way. No single fish would contain sufficient poison to kill or even seriously harm such a bird, but the cumulative effect of a large number of mercury-containing fishes would be deadly, especially as the compound is not readily excreted. Even if a bird were not killed, its reproductive abilities would be impaired. Seed-eating birds are also victims of methylmercury poisoning. Until recently, methylmercuric dicyanadiamide ("Panogen") was the active ingredient in a coating preparation intended to protect seeds from fungal attack. Birds that ate such seeds got poisoned; numerous incidents of this type occurred in Sweden during the 1950's and 1960's, until the use of such coatings was banned. Such seeds, intended for planting, occasionally got ground into flour and eaten. Numerous cases of human methylmercury poisoning occurred in this way.

12.4 ORGANOMETAL(LOID) TOXICITY TOWARDS TERRESTRIAL INVERTEBRATES

Most work involving organometal(loid) toxicity towards land-dwelling invertebrates has concerned the use of these compounds as pesticides. Insects have been the primary target of such work. Organic esters of phosphoric and thiophosphoric acids (eg, Malathion) have long been used for this purpose, and some true organophosphorus compounds (containing a phosphorus-carbon linkage) have also become insecticides. In recent years organotin compounds have made a debut in this role; trade preparations of such materials include Plictran® (tricyclohexyltin hydroxide), Brestan® (triphenyltin acetate), and Du-Ter® (triphenyltin hydroxide). At higher doses, these compounds may prevent the development and maturation of various phytophagous insects, making them useful agricultural tools.

Organotin compounds show strongly biocidal activity against many insect species. They can be used as chemosterilants for the common housefly (Musca domestica) and similar pests. Even at subtoxic levels, these compounds have important effects. When sprayed onto surfaces, organotin compounds frequently repel insects. Such surfaces might include the leaves of plants; in this usage, the compounds are termed antifeedants and are important in contemporary agriculture for the protection of food crops from insect predation. The levels required for this repellant function are considerably lower than those needed for lethal or sterilant effects. Another type of surface that might be protected is skin, human or animal. Certain organosilicon compounds have shown mosquito-repellant action.

Next to insects (mites have been the organisms most studied for biocidal action by organometal(loid)s. Mites feed on the sap of plants and cause extensive damage to crops. The genus Tetranychus contains most of the mite species of concern, and T. urticae (the two-spotted spider mite) is the single most-investigated species. Plictran and related triorganotin compounds act against these mites on plants. They can even destroy mites that infest honeybees without harming the latter, making them especially promising in horticulture.

Other terrestrial invertebrates have been relatively little studied, and then mostly in laboratory work. Scattered reports have appeared involving the effects of organometal(loid)s on earthworms, nematodes, moths, etc. Research involving terrestrial invertebrates has had a very applied focus emphasizing organisms considered detrimental to human interests.

12.5 ORGANOMETAL(LOID) TOXICITY TOWARDS AQUATIC INVERTEBRATES

Research concerning the effects of organometal(loid)s on water-dwelling invertebrates has emphasized two aspects: first, the control of populations of organisms considered "pests"; second, the effects of these compounds on organisms that are important parts of food chains. This second aspect will be considered in the next chapter.

The major target organisms have been freshwater snails, especially the species Biomphalaria glabrata. Such snails serve as intermediate hosts in the life cycles for parasitic flukes of genus Schistosoma. These parasites infect humans causing an affliction called schistosomiasis (snail fever, bilharzia), estimated to involve some two hundred-plus million people throughout the tropical areas of the world.[6,7] For reasons not yet understood, these freshwater snails show an exceptional susceptibility to organotin compounds, especially tri-n-butyltin derivatives. Levels of these compounds which virtually wipe out a complete snail population have little or no effect on fishes. Controlling the populations of these snails has been an important part of a multi-pronged attack on the problem of schistosomiasis. One requirement for the optimal use of these compounds is that their concentration be maintained at an appropriate level—not too high to be toxic to fishes or other vertebrates, but high enough to be effective against the snails. This requirement is met through the use of a "controlled release" formulation,[8,9] usually a paint or a polymeric material in continual contact with the body of water containing the snails. The biologically active organometal is leached out from its matrix at a slow and reasonably constant rate that keeps the dissolved organometal at the appropriate concentration.

Investigations involving other aquatic invertebrates have concentrated on their roles in food chains and the ecological distribution of organometals. Some species, most notably brine shrimp (Artemia salina) and water fleas (Daphnia magna), have been used as test organisms in the determination of relative toxicities of series of organometals. Zoeae of the mud crab Rhithropanopeus harrisii were used to determine structure-activity relationships for the toxicity of di- and triorganotin compounds;[10] the toxicity increased with increasing hydrophobicity. The larvae of the amphipod Gammarus oceanicus showed varying susceptibility to dissolve tri-n-butyltin fluoride at 300 ng/L, whether from the pure compound or as leachate from paint.[11] When rice paddies were treated with triphenyltin hydroxide at a concentration of 1.12 kg per hectare, all aquatic organisms perished. Mosquito larvae populations began to recover after five days, but the populations of predators took longer to recover.[12]

12.6 ANTIFOULING PREPARATIONS CONTAINING ORGANOMETALS

Organometal(loid)s have been incorporated into a variety of controlled release formulations for various uses, such as snail control mentioned in the preceding section. The most extensive such usage has been a variety of antifouling preparations which have become extensive and commercially important.[1,3,13,14] The purpose of these preparations is to protect surfaces in contact with water from being infested with organisms (barnacles, seaweed, etc.) that would settle on them. Such surfaces include the hulls of ships, the underwater sections of wharves and piers, water-carrying pipes, fishermen's nets, etc. In the case of boats and ships, large populations of fouling organisms would greatly hamper the movement and maneuverability of the crafts and add appreciably to fuel costs. Antifouling preparations have been known and used for many years, but organometallic compounds have only relatively recently appeared in this role. Tri-n-butyltin compounds have been most frequently used. For nonporous surfaces, the preparations are usually applied as paints. These slowly release, at a controlled rate, the organotin compound into the surrounding water. A very thin layer forms and remains immediately adjacent to the surface of the paint; this layer repels or kills any free-swimming larvae of barnacles or other sessile organisms that might otherwise attach to the surface. For moving surfaces, this layer rapidly breaks up and must be replenished continuously.

Porous surfaces, such as wood, require somewhat different antifouling preparations. These materials must be protected in their entirety, not merely on the surface. Wood especially is vulnerable to a wide variety of organisms.

Preparations containing tri-n-butyltin compounds are still used, but these will be preparations that can permeate the material to be protected. Mortar and cement readily absorb water; when damp, they can be attacked by molds. Incorporation of tri-n-butyltin compounds into the ingredients prevents this. Numerous related applications have been reported, and continue to appear, in the scientific literature.

12.7 ORGANOMETAL(LOID) TOXICITY TOWARDS MICROORGANISMS

Under contemporary conditions of intensive agriculture, food plants are particularly susceptible to attack by fungus. Rice (Oryzia sativa), a grain which provides the staple diet for a substantial portion of the world's population, has a number of fungal predators, and various organometals have been used or proposed to provide protection. Among these are various derivatives of arsenic, mercury and tin. Wheat and other grains can also be protected in this manner. Sugar beets (Beta vulgaris) are frequently afflicted by a fungal infection named "leaf spot" (Cercospora beticola). Preparations containing triphenyltin compounds have been successfully used to protect sugar beets from this infection. Numerous other food plants have been protected against fungi in similar fashion.[3]

One problem in using organometal(loid)s or other chemicals against fungi (and other microorganisms as well), is that resistant strains almost invariably develop. Excessive use of triphenyltin compounds for sugar beet protection caused tin-resistant strains of leaf spot to develop. Similarly the use of organomercurial preparations such as Panogen® to protect seeds led to the development of resistant strains. Such preparations also led to methylmercury poisoning by birds, mammals and humans who ate the seeds. As a result the use of organomercurials as seed protectants has largely been discontinued. Fungitoxic organometal(loid)s are sometimes applied to soils rather than directly onto plants. This usually happens when the fungus in question attacks plant roots (rather than leaves or stems) or when the plant develops primarily underground.

Bacteria have long been investigated for their susceptibility towards organometal(loid)s, primarily for medicinal purposes (see Chapter 11). More recent studies have concentrated on mechanisms of action and the ways in which organometal(loid)s are metabolized. Organophosphorus compounds frequently prevent or delay bacterial cell wall formation. Other mechanisms of toxicity discussed in Chapter 12.2 apply to microorganisms as well. Bacteria play an important part in the metabolism of environmental organometal(loid)s, as covered in Chapter 13. Vitamin B_{12} is formed by bacteria under environmental conditions, and a variety of alkylphosphonic acids have been isolated from soils, where they apparently serve as natural bactericides. The herbicide Glyphosate is metabolized by a species of Pseudomonas to form glycine,[15] whereas other soil microbes formed amino-methylphosphonic acid instead.

12.8 ORGANOMETAL(LOID)S AND PLANTS

12.8.1 Plant Growth Regulators

The intensive cultivation characteristic of contemporary agriculture, in addition to requiring the use of chemicals for protection against predatory mammals, birds, insects or microorganisms, frequently needs them to control the rate of growth and/or ripening of the plants being cultivated. Various organometal(loid)s have been successfully used for this purpose. By far the most important of these compounds is 2-chloroethylphosphonic acid (Ethephon), which has a variety of trade names. Ethephon hydrolyzes to release ethylene:

$$ClCH_2CH_2PO_3H_2 + H_2O \longrightarrow C_2H_4 + HCl_{(aq)} + H_3PO_{4(aq)} \quad (12.3)$$

The released ethylene is the active agent, and ethylene has been sprayed onto plants for many years to abet their growth. Ethephon has the advantage of being more easily handled and releasing ethylene at a slower rate in a narrower focus. Ethephon has been used to accelerate the ripening of fruit and to ensure a uniformity in the quality of fruit which forms. It has even been reported to induce seedless fruit in grapes, peaches, apples and pears.[16] Ethephon has also been used on vegetables and occasionally on non-food plants, such as cotton, for uniformity of harvest. The effect appears to be concentration-dependent,[3] and the best growth-enhancement seems to occur at levels of about 0.1% (1000 ppm). Related compounds, such as Alsol (Figure 12.1), have also been used for this purpose and act by the same mechanism. Various tetraorganophosphonium salts, especially those having three n-butyl groups, such as Phosphon D (Figure 12.1), also act as growth-enhancing agents. Scattered reports have appeared indicating that other organometalloids may likewise serve this purpose.

12.8.2 Herbicidal Organometalloids

A nebulous and uncertain line exists between the use of an organometalloid as an agent for growth enhancement and the same compound as a growth retardant or herbicide. Sometimes the concentration makes the difference, with enhancement occurring at low levels and retardation at higher levels. The most widely used organometalloidal growth-retardant is N-phosphono-methylglycine (Glyphosate). The related compound, N,N-bis-(phosphono-methyl)glycine(Glyphosine; Figure 12.1), has been used as a ripening agent for sugar cane.[17] Glyphosate occurs in a variety of commercial preparations, the most common being the isopropylammonium salt marketed under the

$ClCH_2CH_2P(:O)(OH)_2$ $ClCH_2CH_2Si(OCH_2CH_2OCH_3)_3$

Ethephon   **Alsol**

$$HOCCH_2NHCH_2P(:O)(OH)_2$$
Glyphosate

$$HOCCH_2N[CH_2P(:O)(OH)_2]_2$$
Glyphosine

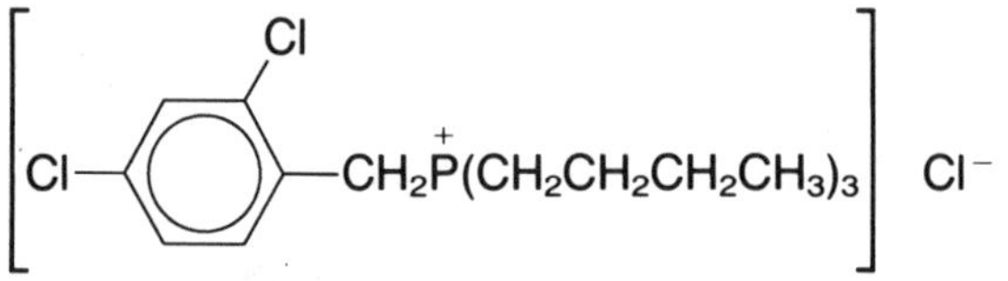

Phosphon D

$CH_3As(:O)(OH)(ONa)$ $(CH_3)_2As(:O)OH$

Methylarsonic Acid,   **Cacodylic Acid**
monosodium salt

Figure 12.1. Organometalloids used as plant-growth control agents.

trade name Roundup®. Glyphosate acts against a wide variety of broad-leafed weeds, apparently by causing foliar necrosis, interfering with plant synthesis of aromatic amino acid, and by inhibiting the enzyme 5-enolpy-ruvylshikimate-3-phosphate synthase.[18] It is commonly used in nurseries and "turf farms" to prevent the growth of weeds in the stock plants.

Another widely used organometalloidal herbicide is methylarsonic acid, usually as its monosodium salt. This compound is commonly employed for the control of weeds in lawns and fields, although it has been used for weed control involving food crops. Sodium cacodylate also serves this purpose, although less commonly. Both methylarsenicals form through biochemical processes (see Chapter 13), and are less toxic to humans than inorganic arsenites. They also bind strongly to soils, thereby becoming inactivated.

12.8.3 Investigations Involving Plants and Organometal(loid)s

Numerous laboratory and field studies concerning the interactions between organometal(loid)s and plants have been reported. The great majority of these have been pragmatic investigations concerning the effects of such compounds as either growth agents or herbicides, but some have dealt with the mechanisms of action. Seed germination and root development are affected by the presence of organometal(loid)s: organometals usually have a strongly inhibitory effect, while organometalloids either have no effect at all or stimulate these processes. Again concentration is a crucial factor, with the probability of deleterious effects increasing at higher concentrations. Glyphosate and other organophosphorus compounds have been used to study plant translocation (movement from soil to plant or through the plants themselves). Phenylmercuric acetate, among others, has been used to study transpiration in plants leaves.[19]

REFERENCES

1. Evans, C J.; Karpel, S., "Organotin Compounds in Modern Technology," Elsevier: Amsterdam, 1985.
2. Peters, R. A.; Stickton, L. A.; Thompson, R. H. S., *Nature,* **1945,** *156,* 616.
3. Thayer, J. S., "Organometallic Compounds and Living Organisms," Academic Press: New York, 1984.
4. Kaegi, J. H. R.; Vallee, B. L., *J. Biol. Chem.,* **1961,** *235,* 2435.
5. Chang, L. W.; Annau, Z., *Neurobehav. Teratol.,* **1984,** 405.
6. Malek, E. A., "Snail-Transmitted Parasitic Diseases," CRC Press: Boca Raton, Florida, 1980.
7. Abdel-Wahab, M. F. (ed), "Schistosomiasis in Egypt," CRC Press: Boca Raton, Florida, 1982.
8. Cardarelli, N. F., "Controlled Release Pesticide Formulations," CRC Press: Cleveland, Ohio, 1976.
9. Baker, R. (ed), "Controlled Release of Bioactive Materials," Academic Press: New York, 1980.
10. Laughlin, R. B.; Johannesen, R. B.; French, W.; Guard, H.; Brinckman, F. E., *Environ. Toxicol. Chem.,* **1985,** *4,* 343.
11. Laughlin, R. B.; Nordlund, K.; Linden, O., *Mar. Environ. Res.,* **1984,** *12,* 243.
12. Schaefer, C. H.; Miura, T.; Dupras, E. F.; Wilder, W. H., *J. Econ. Entomol.,* **1981,** *74,* 597.
13. Thayer, J. S., *J. Chem. Educ.,* **1981,** *58,* 764.
14. Carraher, C. E.; Sheats, J. E.; Pittman, C. U. (eds), "Organometallic Polymers," Academic Press: New York, 1978, pp. 175–218.
15. Jacob, G. S.; Schaefer, J.; Stejskal, E. O.; McKay, R. A., *J. Biol. Chem.,* **1985,** *260,* 5899.
16. Anon., *Chem. Abstr.,* **1984,** *101,* 165577k.
17. Jeffcoat, B., *Chem. Br.,* (June, 1984), 530.
18. Steinruecken, H. C.; Amrhein, N., *Eur. J. Biochem.,* **1984,** *143,* 351.
19. Rahangdale, H. N.; Bhapkar, D. G.; Mungse, H. B., *Chem. Abstr.,* **1984,** *101,* 145951z.

Chapter 13

Organometallic Compounds in Biology: III: Environmental Occurrence and Transformation

13.1 INTRODUCTION

Organometal(loid)s occur in the natural environment to a much greater extent than most people realize.[1] Two routes of entry exist: introduction by humans, either inadvertently or deliberately; and formation from chemical precursors through biological processes or by abiotic chemical reactions.

Organometallic compounds which are deliberately introduced are usually biocidal materials of the types discussed in the preceding chapter. Organoarsenicals[2] and organotin compounds[3] are most likely to be encountered in environmental samplings, although silicones are being reported with increasing frequency. Some organometals have entered environmental waters through industrial wastes, methylmercuric compounds being the best-known example. Tetraalkyllead compounds in gasolines always escape to some extent into the atmosphere.

Methyl derivatives of many elements (even excluding the purely organic derivatives of carbon, nitrogen and oxygen), occur in nature, as indicated in Figure 13.1. While some may be strictly anthropogenic materials, the majority form through processes of chemical or biological methylation. The number and diversity of these methylmetal compounds discovered will increase as analytical techniques become more sensitive and as a wider range of samples are taken. Nor are methylmetal(loid)s the only type of organometal(loid) generated through biological processes; elements such as arsenic, phosphorus, selenium, and cobalt can form bonds to the carbon atoms of other alkyl groups as well.

Studies of environmental organometal(loid)s include investigation into their movement. In addition to purely physical methods of compound transportation (moving water, volatilization, air currents, etc.), natural food

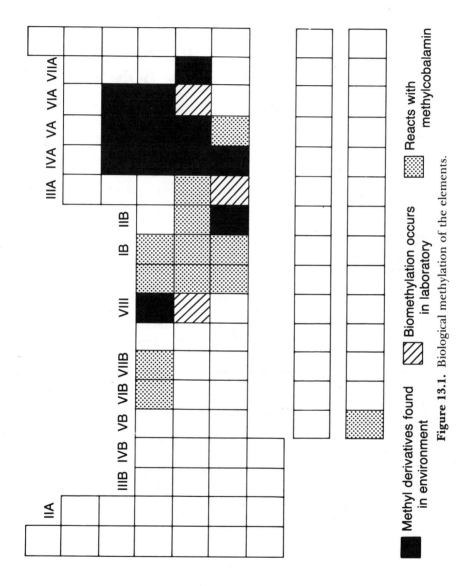

Figure 13.1. Biological methylation of the elements.

Legend:
- Methyl derivatives found in environment
- Biomethylation occurs in laboratory
- Reacts with methylcobalamin

chains also participate in the mobility of such compounds, often to the detriment of the organisms involved. The existence of organometal(loid)s in nature, whether from natural or man-made sources, provides additional pathways for distribution through the biogeochemical cycling of the elements involved.

13.2 BIOLOGICAL FORMATION OF METAL(LOID)–CARBON BONDS

13.2.1 General

The term "biological alkylation" (often contracted to "bioalkylation") refers to various processes by which alkyl groups are introduced onto metal(loid) atoms through the intermediacy of a living organism. While from a purely chemical point of view the reaction is the same inside or outside a cell, the presence of a cell usually drastically alters crucial reaction parameters such as the rate. The most common alkyl group transferred under such conditions is the methyl group, and the term "biological methylation" (or "biomethylation") is frequently mentioned. At the present time there is no evidence for corresponding "bioarylation."

13.2.2 Biological Methylation

The processes and extent of biological methylation have been investigated by many workers, resulting in a substantial literature.[4,5] As Figure 13.1 shows, most of the elements that undergo biomethylation are the heavier representative metals and metalloids. These are the elements whose methyl derivatives are stable towards water. Two compounds appear to be the biological sources for the methyl group, S-adenosylmethionine, **1**, and methylcobalamin (a derivative of Vitamin B_{12}).

Arsenic, among all the elements, has been most investigated for its ability to undergo biomethylation. This grew out of the work on "Gosiogas" (see Chapter 1) by Challenger, who identified it as trimethylarsine, $(CH_3)_3As$, and showed that it formed through the action of the mold Scopulariopsis brevicaulis on arsenious acid present on wallpaper. The methylating agent is S-adenosylmethionine, which has an extensive biochemistry.[6] Methyl groups are introduced one at a time; the first step might be represented by the simplified equation:

$$: AsO(OH)_2^- + RR'SCH_3^+ \longrightarrow CH_3As(:O)(OH)_2 + RR'S \quad (13.1)$$

where $RR'SCH_3^+$ represents S-adenosylmethionine. Methylarsonic acid will in turn be converted to dimethylarsinic acid (cacodylic acid), $(CH_3)_2AsO_2H$, with trimethylarsine representing the third methylation. Most organisms, from bacteria up to and including humans, can methylate arsenic(III) oxide and related oxy anions. This is apparently a mechanism of detoxification, as the methylarsenicals are less toxic than arsenic oxides. Methylarsonic and cacodylic acids are both water-soluble. When formed in mammals, they accumulate in urine and are excreted therein. Industrial workers exposed to arsenic-containing minerals often have elevated arsenic levels in their urine. Methylarsenicals are ubiquitous in natural waters and in most soils. While some of this may arise from the use of such compounds as herbicides, the great majority of naturally occurring methylarsenicals form through biological methylation.

The biomethylation of mercury has been investigated to a degree second only to arsenic. Methylcobalamin serves as the source of the methyl group, and two products can form: salts of the methylmercuric ion, CH_3Hg^+ (collectively termed "methylmercury"), and gaseous dimethylmercury, $(CH_3)_2Hg$. The former usually predominates, but the product ratio depends upon specific conditions. As far as is presently known, biomethylation of mercury occurs almost exclusively in microorganisms, either by bacteria dwelling in the sediments at the bottoms of rivers, lakes, etc., or by microorganisms in the digestive systems of higher organisms. The methyl group transfers, in effect, as an anion to the electrophilic mercury(II) species; ease of transfer apparently varies substantially, depending on the ligands already bonded to mercury. Methylmercurials are not as widespread as methylarsenicals, occurring predominantly in natural waters.

The biomethylation of other elements has received less attention. Selenium and tellurium were included in Challenger's original investigations. Subsequent work on selenium (which has an extensive biochemistry generally similar to that of sulfur) indicated that this metalloid undergoes methylation by essentially the same mechanism as arsenic. Trimethylselenonium ion has been detected in human urine, and dimethylselenide occurs occasionally in the atmosphere. In recent years the formation of methyltin compounds under environmental conditions has generated some concern, since methyltin compounds are toxic to humans and organotins are continuously being introduced into nature through their use as pesticides. Biomethylation of tin apparently occurs by the same mechanism as mercury, and methyltin compounds are usually found in sediments or natural waters. The fact that tin can form up to four tin–carbon bonds per tin atom, plus the fact that these bonds are labile, makes investigation into the behavior of such compounds difficult. The question of whether lead compounds undergo biomethylation has generated controversy. Conflicting evidence has been presented, and the final decision has not yet been reached. Phosphorus, at least in the case of bialaphos (next section), undergoes biomethylation. Methylgermanium and methylantimony compounds have been found in natural waters, and therefore need biomethylation to be formed. Certain derivatives of thallium and rhodium undergo biomethylation under laboratory conditions, but no methyl derivatives of these metals have been found in nature.

13.2.3 Biological Alkylation

Transfer of alkyl groups (including methyl) to nitrogen, oxygen or sulfur is an important part of organismal metabolism.[7] Alkylating agents are receiving considerable research attention for their role in cancer chemotherapy.[8] Introduction of alkyl groups other than methyl onto metals and metalloids has been reported in recent years. Bioalkylation occurs commonly for arsenic, somewhat less commonly for phosphorus and selenium, and rather infrequently for other elements.

Arsenic bioalkylation occurs predominantly in marine organisms, and has been reported for many genera of invertebrates and some vertebrates. As with biomethylation this process appears to be a mechanism for detoxification: highly toxic arsenites or arsenates are converted to much less harmful alkylarsenicals, which are usually sequestered in tissues. The most commonly mentioned compound in this category is arsenobetaine, **2**, which has been recovered from a wide variety of marine fishes, mollusks, crustaceans and plants, as well as a few species of freshwater fishes. Human volunteers or laboratory test organisms who ate fish tissue containing this compound excreted it unaltered and virtually *in toto*. The related compound arsenocholine, **3**, has also been isolated from various marine organisms, as had the arsenolipd **4** and the arseno sugar **5**. All of these compounds have methyl groups attached to arsenic as well as the alkyl group; available evidence suggests that the methyl groups are introduced first, but this has not been firmly established and may depend on the specific compound. This area has received much research attention in recent years, and doubtlessly new compounds of this type will be isolated in years to come.

$$(CH_3)_3\overset{+}{As}CH_2CO_2^-$$

2

$$(CH_3)_3\overset{+}{As}CH_2CH_2OH$$

3

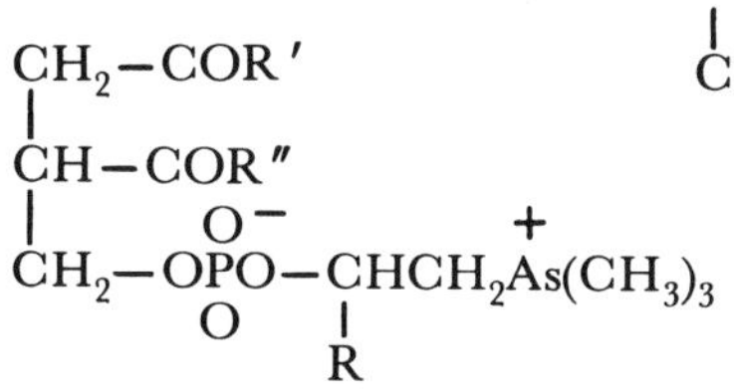

$$\begin{array}{l} CH_2-COR' \\ | \\ CH-COR'' \\ | \qquad\qquad O^- \\ CH_2-OPO-CHCH_2As(CH_3)_3 \\ \qquad\quad O \quad | \\ \qquad\qquad\quad R \end{array} \qquad +$$

$(R=-H,-CO_2H)$

4

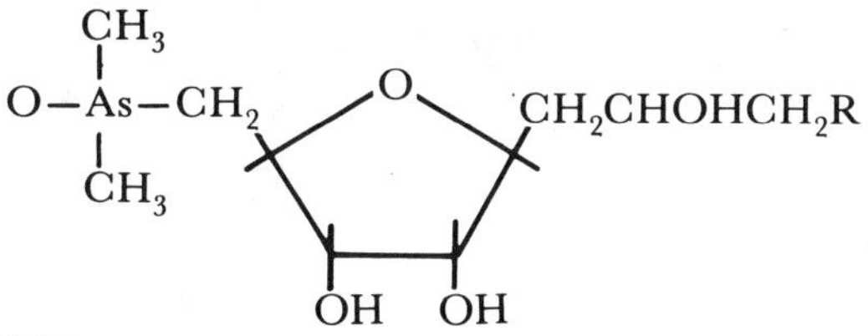

(Arsenic-containing cyclic structure:)

$O-As-CH_2$ with CH_3 groups, ring with O, $CH_2CHOHCH_2R$, OH OH

$(R=-OH,-SO_3H)$

5

Alkylphosphorus compounds occur more extensively in the environment than is generally realized. Phosphonomycin (see Chapter 11) and various related alkylphosphonic acids have been discovered and isolated from soils, where they are formed by fungi as antibacterial agents. A particularly interesting example of this type of compound is bialaphos, **6**, which shows both herbicidal and bactericidal properties. Numerous species of marine organisms generate the compound 2-aminoethylphosphonic acid (ciliatine), **7**. This compound often occurs bound to phosphonolipids,[9] and, in laboratory experiment, has been formed from ^{32}P-labeled phosphoric acid.[10] One mechanism proposes that ciliatine forms by a rearrangement of phosphonol pyruvate;[9] evidence for this mechanism remains indirect and inconclusive.

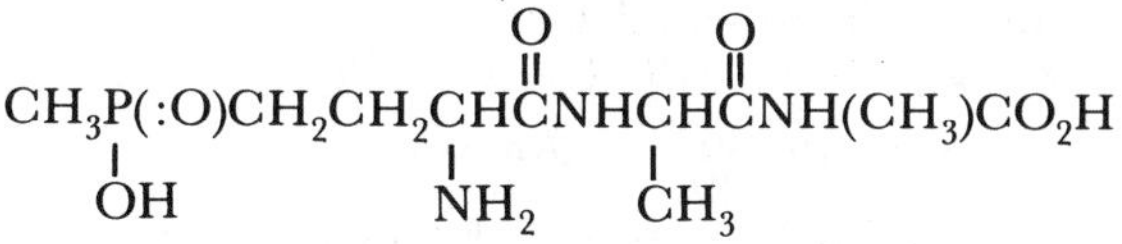

$$\begin{array}{ccc} & O & O \\ & \| & \| \\ CH_3P(:O)CH_2CH_2CHCNHCHCNH(CH_3)CO_2H \\ | & | & | \\ OH & NH_2 & CH_3 \end{array}$$

6

$$H_2NCH_2CH_2P(:O)(OH)_2$$

7

Selenium occurs in many organisms as analogs of sulfur-containing comounds, such as selenomethionine or selenocysteine; quite probably they form by similar mechanisms to the sulfur counterparts. The coenzyme form of Vitamin B_{12} (Chapter 11.6.1) contains an alkyl group bonded to cobalt; in fact, most of its biological importance arises from the ready formation and breaking of cobalt–carbon bonds. As research continues, quite likely the scope of bioalkylation will expand considerably.

13.3 ENVIRONMENTAL CLEAVAGE AND REDISTRIBUTION OF METAL–CARBON BONDS

13.3.1 Biological Reactions

Various species of bacteria possess the ability to cleave mercury–carbon bonds. They convert CH_3HgCl or $(CH_3)_2Hg$ to methane and elemental mercury (which has an appreciable vapor pressure at ordinary temperatures). Phenylmercurials, under the same conditions, yield benzene and mercury. Tetraalkyllead compounds lose one alkyl group to form trialkyllead salts which are water- and lipid-soluble. Further dealkylation may also occur. The less reactive tetraalkyltin compounds tend to lose only one alkyl group to bacterial action. Certain bacterial species, when necessary, can utilize organophosphonic acids as a source of phosphorus by cleaving the phosphorus-carbon bond. Such cleavage, at least for organomercurials, appears to be another method of detoxification. Doubtlessly, more examples of such metal–carbon bond cleavages will be reported as research in this area expands.

13.3.2 Chemical Redistribution in the Environment

Metal–carbon bonds for heavy metals are quite labile, and can undergo facile exchange without the necessity for any biological intervention. Such exchange reactions may result in the effective removal of organometals from some specific environmental location, and/or they might have the opposite effect. Mercury(II) salts have very high demethylating ability, and may form through reactions such as the following:

$$HgCl_{2(aq)} + (CH_3)_3SnCl_{(aq)} \longrightarrow CH_3HgCl_{(aq)} + (CH_3)_2SnCl_2 \quad (13.2)$$

The role that such redistribution reactions play in the environmental formation and decomposition of organometals still remains to be determined.

Metal–carbon bonds also undergo photolysis. Airborne compounds, such as tetraethyllead, are destroyed in the presence of sunlight, and the metal falls to earth. In water such photolysis can occur also, especially on or near the surface, but is usually less complete. Occasionally, as when aqueous mercuric acetate undergoes photolysis, methyl-metal bonds may form.

Soils and sediments affect the redistribution of organometal(loid)s, although the exact extent has not yet been determined. They can bind such compounds, with the extent of binding inversely proportional to the number of organic groups on the metal. Methylarsenicals used as herbicides adsorb onto soil particles, and thereby are moved through the environment or into plant systems. Methylation of mercuric salts by sediments is not solely biological; there are abiotic methylation processes as well. In anaerobic sediments, hydrogen sulfide and its salts are always present and may facilitate the redistribution:

$$2\ CH_3HgCl\ +\ S^=\ \longrightarrow\ (CH_3Hg)_2S\ \longrightarrow$$
$$(CH_3)_2Hg\ +\ HgS\ +\ Cl^-\quad (13.3)$$

The corresponding reaction with selenide has been used as a detoxification process (see Chapter 12.2). Similar resdistribution reactions occur for trimethyltin and trimethyllead sulfides. Nor is there any reason to suppose that such reactions are limited to methyl compounds. A variety of n-butyltin compounds, arising from tri-n-butyltin compounds introduced as pesticides, have been discovered in sediments from various harbors throughout the world.

13.4 MOVEMENT OF ORGANOMETAL(LOID)S THROUGH THE ENVIRONMENT

13.4.1 General

Introduction of one or more organic groups onto a metal or metalloid atom alters the physical properties of the resulting derivative, particularly the solubility and the volatility. Often this may be due to the replacement of bridging oxygen atoms or hydrophilic hydroxyl groups by nonbridging, hydrophobic organic groups. A metal atom in a lattice will have its links to other atoms weakened upon introduction of an organic group, thereby making it easier to detach completely. Solubility in water or hydrocarbons depends very much on the nature and number of organic groups attached; in fact, such solubilities may be related to the calculated molar surface area.[11] Solubility in lipids (hydrocarbons) enables any organometal(loid) to be absorbed by a living organism, which if the compound be not fatal, may transport it elsewhere for eventual excretion.

When all valences of a metal or metalloid are utilized to bond to organic groups through carbon, the resulting molecule has very little tendency towards association, and generally have low melting and boiling points. The best known examples are the "permethylmetals," such as $(CH_3)_2Hg$, $(CH_3)_3Ga$, $(CH_3)_4Sn$, etc., which are volatile liquids or gases at room temperature and are virtually totally insoluble in water but quite soluble in hydrocarbons.

13.4.2 Biological Magnification through Food Chains

Every organism in nature is part of one or more food chains. If a pollutant is introduced into the bottom link of a food chain, its concentration will steadily increase in other parts of that chain and may become fatally large in organisms at the top of the chain. This process is termed "biological magnification" (or biomagnification). The magnification may cause increases in the concentrations of such a pollutant that are hundreds or thousands of times higher than are found in the surrounding environment. Organisms at the top of food chains are usually predators, and such species are the most vulnerable to poisoning by biologically magnified toxins.

Methylmercuric compounds have been the most investigated of organometals in terms of its potential for biomagnification. Such investigations have involved organisms in the field or, increasingly commonly, use model food chains set up and monitored in the laboratory. The most commonly reported model food chain involves four species:

Chlorella	*Daphnia*	*Gambusia*	*Salmo*
vulgaris $\longrightarrow$	*magna* $\longrightarrow$	*affinis* $\longrightarrow$	*gairdneri*
(alga)	(copepod)	(fish)	(rainbow trout)

The majority of food chains studied involve aquatic organisms exclusively, and this seems to be where biomagnification predominates. Food chains involving both aquatic and terrestrial organisms, or terrestrial organisms exclusively are less common.

Methylmercurials undergo biomagnification in a variety of food chains, often showing concentration enhancements of up to 2000-fold. Organoarsenicals by contrast do not show biomagnification. Methyl–arsonic and cacodylic acids are readily excreted in urine due to their solubility in water; trimethylarsine escapes due to its volatility. The alkylarsenicals discussed in Section 13.2.3 are bound tightly to tissues. If they are ingested by another organism (such as humans eating fish containing them), the compounds are excreted unchanged. Preliminary reports suggest that methyltin compounds may undergo biomagnification, but there is no report of any extensive poisonings arising from this, nor have there been any detailed food chain studies. The tendencies of other organometal(loid)s to undergo biomagnification should be predictable by comparing them to mercury or arsenic counterparts.

13.4.3 Environmental Cycles Involving Organometal(loid)s

While environmental cycles involving carbon, nitrogen, phosphorus and sulfur have been known for many years, only recently has the idea that other elements might have similar cycles been generally accepted. Cycles have been proposed for various metals or metalloids, and organo derivatives play an important part in them.[4,5,12,13] Due to their volatility and solubility in water and/or hydrocarbons, organometals can add pathways for the movement of metals. It should be recalled that, while organometal(loid)s are all thermodynamically unstable with respect to oxidation by air, the rates of such oxidation will vary enormously. In consequence, volatile molecules such as $(CH_3)_2Hg$ or $(C_2H_5)_4Pb$ can travel long distances in the atmosphere before they decompose. Metal compounds in sediments can undergo methylation, to be subsequently released into water or atmosphere.

Mercury has been extensively studied, and the role of methylmercurials in the environmental cycling of this element are considerable. Water, air, and food chains all provide pathways for the transport of mercury. While most mercurials in the environment exist as insoluble, nonvolatile species bound in soils or sediments, these are so extensive that even if only a small portion undergoes methylation, enough will escape into water or air to be a problem. Methylmercurials are likely to remain with us for a long time to come.

Arsenic also has its cycle. Volatile species include the previously mentioned trimethylarsine, along with mono- and dimethylarsines, both of which can be formed by fungal action.[4,5] They enable movement through air, while various methylarsenicals may be transported through water, soils or food chains. Much remains to be learned about environmental occurrence of arsenic compounds.

Tin and lead also apparently have their own cycles in nature. Organo derivatives of both metals, especially tin,[2] enter the environment in substantial quantities, and methyltins (probably methylleads also, although this is not certain) form through biomethylation. How these processes will affect the movement of those metals still must be determined.

In principle, every element should have its own environmental cycle. Those elements which form organo derivatives that have stability, even if only for short periods of time under natural conditions should possess cycles in which those derivatives participate. Phosphorus, selenium, tellurium, germanium, antimony and iodine certainly fall into this category, with silicon being an outside possibility; there may be other elements currently unknown and unsuspected. As the formidable experimental difficulties disappear, environmental cycles will become better and better known, and the role of organometal(loid)s in such cycles may well be considerably more important than presently is realized or even suspected.

REFERENCES

1. Craig, P.J., in "Comprehensive Organometallic Chemistry," Pergamon: London, 1982, Volume 2, pp. 979–1020.
2. Fowler, B. A., ed., "Biological and Environmental Effects of Arsenic," Elsevier: Amsterdam, 1983.
3. Blunden, S. J.; Hobbs, L. A.; Smith, P. J., in "Environmental Chemistry," (Bowen, H. J. M., ed.), Periodical Reports, The Royal Society of Chemistry: London, 1984, pp. 49–77.
4. Thayer, J. S., "Organometallic Compounds and Living Organisms," Academic Press: New York, 1984, pp. 189–245.
5. Thayer, J. S.; Brinckman, F. E., *Adv. Organomet. Chem.*, **1982**, *20*, 313.
6. Salvatore, F.; Borek, E.; Zappia, V.; Williams-Ashman, H. G.; Schlenck, F., eds., "The Biochemistry of Adenosylmethionine," Columbia University Press: New York, 1977.

7. Paik, W. K.; Kim, S., "Protein Methylation," Wiley-Interscience: New York, 1980.
8. Cline, M. J.; Haskell, C. M., "Cancer Chemotherapy," 3rd ed., Saunders: Philadelphia, 1980, pp. 31–44.
9. Kittredge, J. S.; Roberts, E., *Science,* **1969,** *164,* 37.
10. Miceli, M. V.; Henderson, T. O.; Myers, T. C., *Science,* **1988,** *209,* 1245.
11. Laughlin, R. B.; French, W.; Johannesen, R. B.; Guard, H. E.; Brinckman, F. E., *Chemosphere,* **1984,** *13,* 575.
12. Hutainger, O. ed., "The Handbook of Environmental Chemistry. 1A. The Natural Environment and the Biogeochemical Cycles," Springer-Verlag: Berlin, 1980.
13. Ridley, W. P.; Dizikes, L. J.; Wood, J. M., *Science,* **1977,** *197,* 329.

Chapter 14

Afterword

14.1 INTRODUCTION

During the months since the completion of the preceding chapters, several hundred additional papers on organometal(loid) compounds and their chemistry have been published. A comprehensive review of these papers is not practical; however, certain selected very recent papers are mentioned here to extend material previously mentioned and/or to point out new possibilities in organometal(loid) research.

14.2 SYNTHESES INVOLVING ORGANOMETALLIC COMPOUNDS

The state of metal surfaces has long been recognized as being crucial when such metals are involved in the preparation of organometals and/or as catalysts. Metal powders, upon treatment with ultrasound, show much enhanced reactivity towards organic molecules.[1] This enhancement was attributed to cleansing of the surfaces of impurities. The role of ultrasound (sonochemistry) in preparations and reactions involving organometals has recently been reviewed,[2] and promises to become an exciting new area of investigation.

The first direct metalation of ethylene, giving $CH_2=CHK$, arose from the treatment of ethylene with n-butyllithium, potassium t-butoxide and tetramethylethylenediamine at $-25°$ in hexane.[3] Product distribution in the reaction of boron trifluoride etherate with ethylmagnesium bromide changed markedly with solvent:[4]

$$3.5\ C_2H_5Br + 3.5\ Mg + BF_3\cdot O(C_2H_5)_2 \diagup \diagdown$$

$$\text{(A)}\ (C_2H_5)_3B + 3\ MgBrF + \tfrac{1}{2} C_2H_5MgBr$$

$$\text{(B)}\ \tfrac{7}{8} MgBrB(C_2H_5)_4\cdot nC_4H_8O + 2\tfrac{1}{8} MgFBr + \tfrac{1}{8} BF_3\cdot C_4H_8O \qquad (14.1)$$

Reaction (A) was done in diethyl ether with a yield of 99%. Reaction (B) was done in tetrahydrofuran with a yield of 87.5%. A new iodonium compound was prepared in methylene chloride by the following reaction:[5]

$$C_6H_5IO + (CH_3)_3SiC\equiv CR \xrightarrow[(C_2H_5)_3O^+ \ BF_4^-]{} RC\equiv CIC_6H_5^+ \ BF_4^- \quad (14.2)$$

The role of organoiodine(III) compounds in organic synthesis has been reviewed.[6] Optically active oligomeric helical metallocenes have also been prepared and investigated,[7] as have polymeric compounds containing organogallium(I) units.[8] Complexes containing alkylruthenium(VI) groups, such as $(C_4H_9)_4N^+$ $[(CH_3)_4Ru\equiv N]^-$, have been reported for the first time.[8]

More complexes involving unsaturated organometalloids and metals have been reported, including a ruthenium–diborabenzene species[9] and a bis complex of cobalt to arsabenzene.[10] Metal–boarabenzene complexes have been reviewed,[11] as have carbene/carbyne complexes of ruthenium, osmium and iridium.[12] The unusual compound $L_2Ga^+GaCl_4^-$ (L = 2.2 paracyclophane) contains synergistic gallium–carbon linkages.[13]

14.3 COMPOUNDS INVOLVING METAL–METAL MULTIPLE BONDING

The number of papers on compounds containing multiply-bonded metals or metalloids continues to increase rapidly, and this is an area that continues to draw much attention. The published work on disilenes has reached the point where a review article might appear; one such article has just appeared.[14] 1,4-Disilabenzene was trapped in an argon matrix at 10 K and characterized.[15] The stable molecule $[(CH_3)_2CH]_2Si=NMes$ (Mes = mesityl) has been isolated.[16] A compound with a P(III)—C double bond undergoes photoisomerization:[17]

$$\underset{\text{C}_6\text{H}_5}{\overset{\text{Mes}}{\cdot\cdot P=C}}\overset{\text{H}}{\underset{}{}} \quad \underset{h\nu}{\rightleftarrows} \quad \underset{\text{H}}{\overset{\text{Mes}}{\cdot\cdot P=C}}\overset{\text{C}_6\text{H}_5}{} \quad (14.3)$$

The pale-yellow solid Mes—C≡As (m.p. 114–6°) represents the first stable arsaethyne (As analog of a nitrile) to be isolated.[18] Two papers concerning compounds with B=P linkages have appeared.[19,20] These compounds also contain mesityl groups. As mentioned on page 66, the presence of bulky groups is necessary for the isolation of such compounds, and the mesityl group continues to be the group of choice.

Multiply bonded metals in organometals are also appearing in growing numbers. The molecule $Cl_3WC_3(CH_3)_3$ contains a 4-membered W—C—C—C ring with delocalized pi-bonding.[21] The molecule Mes—$PCo_2(CO)_6$ contains P=Co bonds.[22] The compound $Ar(OC)_2Mn=Ge=Mn(CO)_2Ar$ (Ar = η^5-pentamethylcyclopentadienyl) is the first reported species containing manganese–germanium double bonds.[23] The red crystalline solid $R_2Sn=P$—Mes (R=$(Me_3Si)_2CH^-$) decomposes quite slowly at room temperature.[24] A compound containing both a tungsten-carbon double bond and a tungsten–platinum single bond has been reported.[25] The highly catenated species $Na^+ P_{21}(CH_3)_2$ may contain some P—P double bonding.[26]

14.4 ORGANOMETALLIC COMPOUNDS IN THE ENVIRONMENT

The appearance, movement and role of organometallic compounds in the natural environment continues to attract research attention and concern. A monograph on this topic has just been published.[27] The occurrence and transformation of organometal(loid)s in natural waters has also been the subject of a recent review,[28] as has the role of organolead compounds.[29]

The widespread use of organotin compounds in antifouling preparations (Chapter 12.6) has led to increasing work on their role in the environment, especially in invertebrates used for food (clams, oysters, crabs). There is increasing evidence that such organisms can accumulate organotin compounds. The marine mussel Mytilus edulis can accumulate bis(tri-n-butyltin) oxide,[30] and the freshwater clam Anodonta anatina can accumulate (n-$C_4H_9)_2SnCl_2$, with effects on its metabolism.[31] The effects of leachates from tri-n-butyltin-impregnated surfaces on freshwater mollusks has been reviewed.[32]

Active research on biological methylation in all its aspects continues in many countries. An international conference on this topic was held during September 1987 in London, England, in honor of the centennial of the late Professor Frederick Challenger. In an investigation into the various species of sediment bacteria that might methylate mercury, workers found that the sulfate-reducing species Desulfovibrio desulfuricans show especially vigorous methylating activity.[33] While most mammals have the ability to methylate arsenic, such ability is apparently not universal; the marmoset monkey, Callithrix jacchus, apparently lacked this ability.[34]

Accumulating evidence indicates that arsenobetaine is the most commonly occurring organoarsenical in the tissues of marine food organisms (see Chapter 13.2.3). It has been reported as the predominant compound (ca. 90%) in the red crab,[35] as well as in other crab species[36] and in certain fishes.[37] An investigation of arsenobetaine involving test organisms indicated that this compound was not toxic, nor was it mutagenic.[38]

14.5 CONCLUSION

Chapter 1 opened with the statement "organometallic chemistry has developed into a recognized subarea of chemistry. . . . " It is arguably the most diversified, and, as such, shows some tendency to be partitioned off by other subareas of Chemistry. Even the boundaries of organometallic chemistry are somewhat nebulous. The major purpose of this book has been to act, as much as possible, as a unifying force, bringing together the various parts of this variegated field and indicating something about other relationships. This book has also tried to indicate the dynamism, the surging energy that this field currently is showing, both in fundamental research and in the increasing number of applications to industry and other aspects of daily life. How well it has succeeded will be decided by you, the readers.

REFERENCES

1. Boudjok, P.; Thompson, D. P.; Ohrbom, W. H.; Han, B. H., *Organometallics*, **1986**, *5*, 1257.
2. Suslick, K. S., *Adv. Organomet. Chem.*, **1986**, *25*, 73.
3. Brandsma, L.; Verkruijsse, H. D.; Schade, C.; Schleyer, P. v. R., *J. Chem. Soc., Chem. Comm.*, **1986**, 260.
4. Brown, H. C.; Racherla, U. S., *Organometallics*, **1986**, *5*, 391.
5. Ochiai, M.; Kunishima, M.; Sumi, K.; Nagao, Y.; Fujita, E.; Arimoto, M.; Yamaguchi, H., *Tetrahedron Lett.*, **1985**, *26*, 4501.
6. Moriarty, R. M.; Prakash, O., *Accts. Chem. Res.*, **1986**, *19*, 244.
7. Sudhakar, A.; Katz, T. J.; Yang, B. W., *J. Am. Chem. Soc.*, **1986**, *108*, 2790.
8. Shapley, P.A.; Wepsiec, J. P., *Organometallics*, **1986**, *5*, 1515.
9. Herberich, G. E.; Hessner, B.; Hostalek, M., *Angew. Chem.*, **1986**, *98*, 637.
10. Elschenbroich, C.; Kroker, J.; Massa, W.; Wuensch, M.; Ashe, A. J., *Angew. Chem.*, **1986**, *98*, 652.
11. Herberich, G. E.; Ohst, J., *Adv. Organomet. Chem.*, **1986**, *25*, 199.
12. Gallop, M. A.; Roper, W. R., *Adv. Organomet. Chem.*, **1986**, *25*, 121.
13. Schmidbaur, H.; Bublak, W.; Huber, B.; Mueller, G., *Organometallics*, **1986**, *5*, 1647.
14. Brook, A. G.; Baines, K. M., *Adv. Organomet. Chem.*, **1986**, *25*, 1.
15. Maier, G.; Schoettler, K.; Reisenauer, H. P., *Tetrahedron Lett.*, **1985**, *26*, 4079.
16. Hesse, M.; Klingebiel, U., *Angew. Chem.*, **1986**, *98*, 638.
17. Yoshifuji, M.; Toyota, K.; Inamoto N.; Hirotsu, K.; Higuchi, T., *Tetrahedron Lett.*, **1985**, *26*, 6443.
18. Maerkl, G.; Sejpka, H., *Angew. Chem. Int. Ed.*, **1986**, *25*, 264.
19. Arif, A. M.; Boggs, J. E.; Cowley, A. H.; Lee, J. G.; Pakulski, M.; Power, J. M., *J. Am. Chem. Soc.*, **1986**, *108*, 6083.
20. Bartlett, R. A.; Feng, X.; Power, P. M., *J. Am. Chem. Soc.*, **1986**, *108*, 6817.
21. Latham, I. A.; Sita, L. R.; Schrock, R. R., *Organometallics*, **1986**, *5*, 1508.
22. Arif, A. M.; Cowley, A. H.; Pakulski, M., *J. Chem. Soc., Chem. Comm.*, **1985**, 1707.
23. Korp, J. D.; Bernal, I.; Hoerlein, R.; Serrano, T.; Herrmann, W. A., *Chem. Ber.*, **1985**, *118*, 340.
24. Couret, C.; Escudie, J.; Satge, J.; Raharinirina, A.; Andriamizaka, J. D., *J. Am. Chem. Soc.*, **1985**, *107*, 8280.
25. Elliot, G. P.; Howard, J. A.; Nunn, C. M.; Stone, F. G. A., *J. Chem. Soc., Chem. Comm.*, **1986**, 431.
27. Craig, P. J., "Organometallic Compounds in the Environment," J. Wiley & Sons, New York, 1986.

28. Chau, Y. K., *Sci. Total Environ.*, **1986,** *49*, 305.
29. De Jonghe, W. R. A.; Adams, F. C., *Adv. Environ. Sci. Technol.*, **1986,** *17*, 561.
30. Laughlin, R. B.; French, W.; Guard, H. E., *Environ. Sci. & Technol.*, **1986,** *20*, 884.
31. Holwerda, D. A.; Herwig, H. J., *Bull. Environ. Contam. Toxicol.*, **1986,** *36*, 756.
32. Thain, J. E.; Waldock, M. J., *Water Sci. Technol.*, **1986,** *18*, 193.
33. Compeau, G. C.; Bartha, R., *Appl. Environ. Microbiol.*, **1985,** *50*, 498.
34. Vahter, M.; Marafante, E., *Arch. Toxicol.*, **1985,** *57*, 119.
35. Matsuto, S.; Stockton, R. A.; Irgolic, K. J., *Sci. Total Environ.*, **1986,** *48*, 133.
36. Francesconi, K. A.; Micks, P.; Stockton, R. A.; Irgolic, K. J., *Chemosphere*, **1985,** *14*, 1443.
37. Hanaoka, K.; Tagawa, S., *Biol. Abstr.*, **1985,** *80*, 94162.
38. Jongen, W. M. F.; Cardinaals, J. M.; Bos, P. M. J.; Hagel, P., *Biol. Abstr.*, **1985,** *80*, 109654.

Subject Index

Author Index

Compound Index

*Italicized page number indicates compound appears in a figure.